Primary Maths for Scotland

1st Level Maths
Textbook 1B

Series Editor: Craig Lowther

Authors: Antoinette Irwin, Carol Lyon,
Kirsten Mackay, Felicity Martin, Scott Morrow

© 2019 Leckie

001/31052019

10 9 8 7

ISBN 9780008313968

Published by
Leckie
An imprint of HarperCollins*Publishers*
Westerhill Road, Bishopbriggs, Glasgow, G64 2QT
T: 0844 576 8126   F: 0844 576 8131
leckiescotland@harpercollins.co.uk   www.leckiescotland.co.uk

HarperCollins Publishers
1st Floor, Watermarque Building, Ringsend Road, Dublin 4, Ireland

Publisher: Fiona McGlade
Managing editor: Craig Balfour
Project editors: Alison James and Peter Dennis

Special thanks
Answer checking: Caleb O'Loan
Copy editing: Louise Robb
Cover design: Ink Tank
Layout and illustration: Jouve
Proofreading: Dylan Hamilton

A CIP Catalogue record for this book is available from the British Library.

Acknowledgements

Images © Shutterstock.com

Printed in Italy by Grafica Veneta SpA

Contents

Answers and free downloadable resources

Answers

All answers to the Before we start, Let's practise and Challenge questions in Textbooks 1A, 1B and 1C can be downloaded from our website here:

https://collins.co.uk/primarymathsforscotland

Free downloadable resources

There are free downloadable resources to support Textbooks 1A, 1B and 1C. These can be downloaded, printed out and photocopied for in-class use from our website here:

https://collins.co.uk/primarymathsforscotland

There are two types of resources:

- **General resources.** These are helpful documents that can be used alongside Textbooks 1A, 1B and 1C, and include, for example, blank ten frames, number lines, 100 squares and blank clock faces.

- **Specific resources.** These are supporting worksheets that relate to either a particular area of learning or a specific question and are labelled with a unique resource reference number. For example, 'Resource 1B_2.3_Let's_Practise_Q1' is a specific downloadable resource for Textbook 1B, Chapter 2.3, Let's practise Question 1.

1.1 Rounding

We are learning to round numbers to the nearest ten.

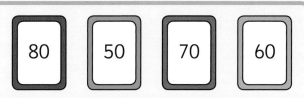

Before we start

Put these decade numbers in order from smallest to biggest.

| 80 | 50 | 70 | 60 |

We can use number lines to help us round up or down.

Let's learn

We can use a number line to help us round numbers to the nearest ten.

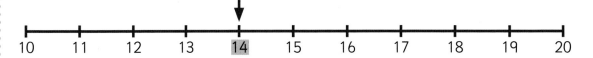

The number 14 is between 10 and 20, but it is closer to the number 10, so we round **down**.

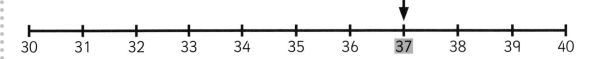

The number 37 is between 30 and 40, but it is closer to the number 40, so we round **up**.

If a number ends in 5, we usually round **up**.

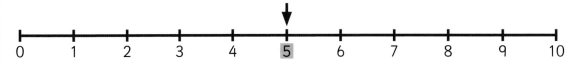

1) Which is the closest ten to these numbers? Use the number lines to help you.

Number lines

a) 33
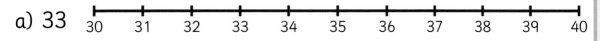

b) 48
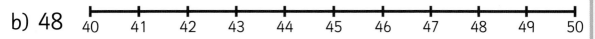

c) 54
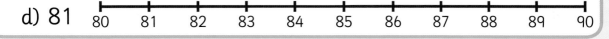

d) 81

2) Amman has tried rounding these numbers to the nearest ten. For each one, say if he is right or wrong and explain why.

Example:

87 should be rounded down to 80.

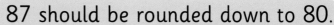

Amman is wrong. 87 is only 3 away from 90. It should be rounded up to 90 as it is closer to 90 than to 80.

54 should be rounded up to 60.

a)
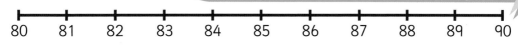

98 should be rounded up to 100.

b)
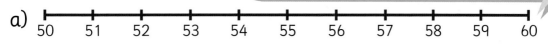

27 should be rounded down to 20.

c)
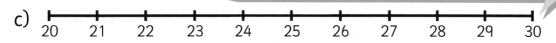

13 should be rounded up to 20.

d)
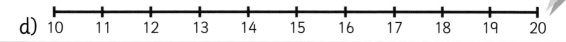

3) Amman says that 95 rounded to the nearest ten is 90. Isla thinks that it is 100.

I think 95 rounded to the nearest ten is 90.

I think 95 rounded to the nearest ten is 100.

Who is right?

Use a number line to help you.

CHALLENGE!

Number lines

In pairs, play the following game:

1) Write down a secret three-digit number (for example, 433).

2) Round the number to the nearest ten and tell your partner what the new number is (in this example it would be 430).

3) Your partner then has three attempts to try and guess the secret number. They get three points if they guess it first time, two points if they guess it second time, and one point if they guess it third time.

4) Play the game several times and swap roles each time. To make things really tough, try rounding the number to the nearest 100!

1.2 Estimating the answer by rounding

We are learning to estimate by rounding.

Before we start

What is the decade number that comes after this number?

| 40 |

We can use rounding to estimate if an answer is correct.

Let's learn

Isla and Amman are working out the answer to 12 + 19. Isla gets the answer 21. Amman thinks the answer is 31.

We can use rounding to estimate the answer.

12 rounded to the nearest ten is 10.

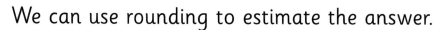

19 rounded to the nearest ten is 20.

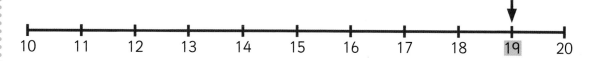

10 + 20 = 30 so we can estimate the answer should be close to 30. We can see Amman's answer of 31 is more likely to be correct.

1) Estimate the answer to these problems by rounding to the nearest ten.

a) 18 + 13

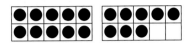

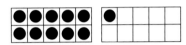

b) 9 + 14

c) 24 + 11

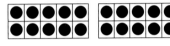

2) Estimate the answer to these problems by rounding to the nearest ten. Use a number line to help you.

a) 28 – 19   b) 32 – 8   c) 37 – 12

Number lines

3) Use rounding to estimate the answer and see who has got the right answer to this problem:

19 + 18

I think the answer is 37.

I think the answer is 27.

Work with a partner.

Student A: Make two piles of beads. Count the number in each pile. Then add the numbers together. Write the addition and the answer but do not show it to your partner.

Student B: Estimate the number of beads in each pile. Write down your estimates. Estimate the total number of beads and write it down.

Student A: Look at your partner's estimate. Help them to find the exact answers by saying: 'The number is less than your estimate' or 'The number is more than your estimate'.

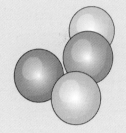

Then swap roles.

2 Number – order and place value

2.1 Reading and writing two-digit numbers

We are learning to read and write two-digit numbers in words.

Before we start

Write down the pairs that **do not** match.

| 3 | three | 80 | eighty | 5 | eight | 1 | one |
| 4 | forty | 90 | nineteen | 15 | fifty | 20 | twenty |

Numbers can be written in numerals and in words.

Let's learn

Do you remember how to spell these number names?

1	2	3	4	5	6	7	8	9
one	two	three	four	five	six	seven	eight	nine

10	20	30	40	50	60	70	80	90
ten	twenty	thirty	forty	fifty	sixty	seventy	eighty	ninety

We can write any number between 20 and 100 in words.

26 join the words twenty and six to make twenty-six

43 join the words forty and three to make forty-three

84 join the words eighty and four to make eighty-four

1) Use a 100 square.

   a) Colour number eighteen **blue**.

   b) Colour number seventy-nine **green**.

   c) Colour number forty **pink**.

   d) Colour number thirty-two yellow.

   e) Colour number ninety-five **purple**.

   f) Colour number sixty-one **orange**.

100 square

2) Write these numbers in words in your jotter.

   a) 49   b) 37   c) 51   d) 72   e) 88

3) Match each balloon to its number name.
   For example, **12** matches with **twelve**.

Resource
1B_2.1_Let's
Practise_Q3

47   14   15   92

21

fifteen   thirty-two

twenty-one

forty-one   ninety-two

12   sixty-seven

twelve   forty-seven

32

fourteen   twenty-eight

67   41   28

CHALLENGE!

Write each missing number in numerals and in words.

a) 35   36   **?**   38   39   40   **?**   b) **?**   86   87   88   89   **?**   91

2.2 Naming and ordering the hundreds

We are learning to name and order the hundreds numbers.

Before we start

Say and write the missing numbers in your jotter.

a)

10	20	?	?	?	?	?	?	?	?

b)

100	90	?	?	?	?	?	?	?	?

We can use the 1 to 9 pattern to help us name the hundreds numbers.

Let's learn

Say the number names out loud.

100	200	300	400	500	600	700	800	900
one hundred	two hundred	three hundred	four hundred	five hundred	six hundred	seven hundred	eight hundred	nine hundred

There are one hundred dots in each square. Count out loud.

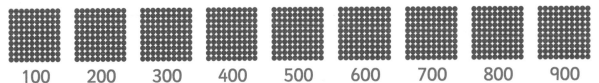

100   200   300   400   500   600   700   800   900

Count **backwards** in hundreds along the number line.

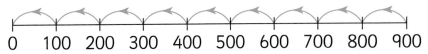

0   100   200   300   400   500   600   700   800   900

1) How many dots?
   Write each answer in words and in numerals in your jotter.

   a)

   b)

   c)

   d)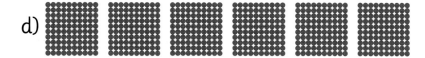

2) Say and write the missing numbers in words and in numerals.

   a)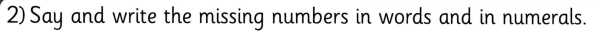
   0   ?   200   300   ?   ?   600   ?   800   ?   1000

   b)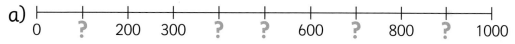
   0   ?   ?   ?   ?   500   ?   ?   ?   ?   1000

   c)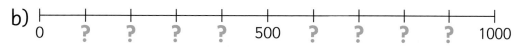
   0   100   ?   300   ?   500   ?   700   ?   900   ?

⭐ **CHALLENGE!** ....

One hundred more than nine hundred is one thousand.
We write **1000** in numerals.

Continue the pattern. How far can you go?

800	**900**	**1000**	**1100**	**1200** ...
eight hundred	nine hundred	one thousand	one thousand, one hundred	one thousand, two hundred...

2.3 Reading and writing three-digit numbers

We are learning to read and write three-digit numbers.

Before we start

Can you find the number words that match these numerals? They are all hidden in the wordsearch.

Resource
1B_2.3_Before_
we_start

72   12   17

30

1

t	h	r	e	e	d	a	l	t	e	n
e	s	o	y	s	i	x	t	e	e	n
o	n	e	h	m	s	e	v	e	n	r
t	w	e	l	v	e	p	f	i	v	e
f	o	r	t	y	-	f	o	u	r	b
s	e	v	e	n	t	e	e	n	a	g
y	t	h	i	r	t	y	v	r	e	d
s	e	v	e	n	t	y	-	t	w	o
s	i	x	t	f	e	l	e	v	e	n
o	n	e	h	u	n	d	r	e	d	c
q	t	w	e	n	t	y	-	o	n	e

10

5

11

21

16

3

7   100   44   6

We can use what we know about writing two-digit numbers and the hundreds numbers to help us to write three-digit numbers.

175 This number says one hundred and seventy-five.
The number 175 has three digits. It is a **three-digit number.**

480 This number says four hundred and eighty.
The number 480 has three digits. It is a **three-digit number.**

Remember to write the word 'and' after the word 'hundred'.

Let's practise

1) Write the missing numerals or number words for each lorry in your jotter.

Resource
1B_2.3_Let's_Practise_Q1

a)
367 — three hundred and sixty-seven

b)
190 — ?

c)
? — two hundred and fifty

d)
147 — ?

e)
? — four hundred

f)
222 — ?

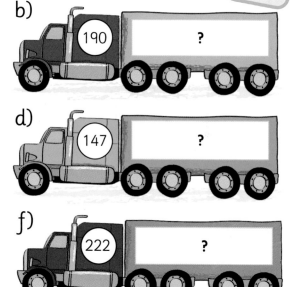

CHALLENGE!

How many different three-digit numbers can you make with these number cards?

Write them down in numerals and in words.

| 6 | 7 | 8 |

2.4 Counting forwards in ones

We are learning to count forwards in ones from any three-digit number.

Before we start

Fill in the missing numbers, counting on by ones. Say the names of the numbers as you count them.

a) 39, **?**, 41, 42, 43, 44, 45, **?**, 47, 48, 49, **?**

b) 68, 69, **?**, 71, 72, 73, **?**, 75

c) 89, **?**, 91, 92, 93, **?**, 95, 96

Patterns can help us to say, read and write numbers in the correct number order.

Let's learn

Read the numbers out loud.
Talk about the patterns you see.

101	102	103	104	105	106	107	108	109	110
111	112	113	114	115	116	117	118	119	120
121	122	123	124	125	126	127	128	129	130
131	132	133	134	135	136	137	138	139	140
141	142	143	144	145	146	147	148	149	150
151	152	153	154	155	156	157	158	159	160
161	162	163	164	165	166	167	168	169	170

Can you keep the pattern going? How far can you go?

Let's practise

1) a) Write the missing door numbers in numerals and in words.
   b) Amman has a blue front door.
      What number door does Amman have?

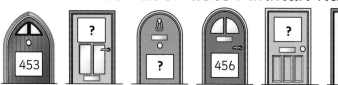

2) Count forwards in ones to find the number on each red card.
   Write the answers in your jotter.

   a) 786 ? ? ? ? ? ?

   b) 658 ? ? ? ? ? ?

   c) 595 ? ? ? ? ? ?

CHALLENGE!

Amman rolls 5, 6 and 3 with numeral dice and writes down the number 563. Then he writes the next eight numbers in the sequence.

Start

| 563 | 564 | 565 | 566 | 567 | 568 | 569 | 570 | 571 |

1) Roll three numeral dice to make your own three-digit number.
   Read your number out loud and write it down.
   Write the next eight numbers in the sequence.

2) Ask a partner to check your answers.
   Challenge them to continue your sequence.
   How far can they go?

2.5 Counting backwards in ones

We are learning to count backwards in ones starting at any three-digit number.

Before we start

Count backwards in ones.
Write the missing numbers in your jotter.

a)
? ? ? ? ? ? 21

b)
? ? ? ? ? 53

Patterns can help us to say, read and write numbers in the correct number order.

Let's learn

Read the numbers in each row **backwards**.

341	342	343	344	345	346	347	348	349	350
331	332	333	334	335	336	337	338	339	340
321	322	323	324	325	326	327	328	329	330
311	312	313	314	315	316	317	318	319	320
301	302	303	304	305	306	307	308	309	310

Isla counts backwards from 301. What do you notice?

301, 300, 299, 298, 297, 296 …

1) Count backwards in ones, from:

a) 444 | 443 | 442 | ? | ? | ?

b) 611 | ? | ? | ? | ? | ?

c) 831 | ? | ? | ? | ? | ?

d) 782 | ? | ? | ? | ? | ?

2) Count backwards in ones to number the bottles:

a)

304 ? ? ? ?

b)

701 ? ? ? ?

c)

500 ? ? ? ?

d)

880 ? ? ? ?

⭐ **CHALLENGE!** ....

Resource 1B_2.5_ Challenge

a) Start at the blue arrow. Count backwards from 401 in ones. Write down the numbers you say on a number line.

➡ 401 | ? | ? | ? | ? | ? | ? | ? | ?

b) Now try this with 301, 201 and 101. What do you notice?

2.6 Before, after and in-between

We are learning to say and write the number before, after and in-between.

Before we start

a) Write down the missing numbers in your jotter.

Resource
1B_2.6_Before_
we_start

? 37 38 39 40 41 42 43 44 ? 46 47 48 49 50 ? 52 53 54 55 56 57 58 59 60 61 62 ? 64 65 66 67 68 ? 70

b) What number comes **after** 70?

c) What number comes **before** 41?

d) Write any three numbers that come **in-between** 50 and 60.

We count forwards to find the number after. We count backwards to find the number before.

Let's learn

Read the numbers on the number line.

193   194   195   196   197   198   199   **200**   201   202   203

The number **after** 200 is 201.

The number **before** 200 is 199.

The number 200 is **in-between** 199 and 201.

The numbers 194, 195, 196, 197, 198, 199, 200, 201 and 202 all come **in-between** 193 and 203.

1) Say the missing numbers out loud. Write them down in your jotter. Read them to a partner.

a) 206 ? ? ? ? ? ? ? ? ? 216

b) 598 ? ? ? ? ? ? ? ? ? 608

c) 103 ? ? ? ? ? ? ? ? ? 113

d) 899 ? ? ? ? ? ? ? ? ? 909

e) 643 ? ? ? ? ? ? ? ? ? 653

2) Write the number that comes **after**:

a) 209   b) 600   c) 399   d) 740
e) 831   f) 110   g) 570   h) 962

3) Write the number that comes **before**:

a) 110   b) 901   c) 470   d) 800
e) 281   f) 599   g) 319   h) 700

4) Write all the numbers that come **in-between**:

a) 408 and 418   b) 797 and 807
c) 326 and 336   d) 600 and 610
e) 555 and 565   f) 200 and 210

5) Complete the following number sequences:

a) 673 | 674 | 675 | 676 | ? | ? | 679 | ?

b) 894 | ? | ? | 891 | ? | 889 | ? | ?

c) ? | ? | ? | 782 | ? | 784 | ? | ?

d) ? | ? | 500 | ? | ? | ? | 504 | ?

⭐ **CHALLENGE!**

a) Make four different three-digit numbers from the digits 2, 9 and 1.

192 | ? | ? | ?

b) Put your chosen numbers in the central box and complete the number sequence on either side.

190	191	192	193	194
a) ?	?	?	?	?
b) ?	?	?	?	?
c) ?	?	?	?	?
d) ?	?	?	?	?

2.7 Counting forwards in tens and hundreds

> We are learning to count forwards in tens and hundreds.

Before we start

Some numbers have fallen off these lockers.

a) Write the missing numbers in numerals in your jotter.

b) The eighth locker in the third row belongs to Finlay. Write this number in words.

395	396	397	?	?	?	?	402
403	404	?	?	?	408	409	?
?	?	413	414	?	?	?	?

> Looking for patterns can help us when counting forwards in tens and hundreds.

Let's learn

Read down each column of numbers. Each number is **ten more** than the number before it.

101	102	103	104	105	106	107	108	109	110
111	112	113	114	115	116	117	118	119	120
121	122	123	124	125	126	127	128	129	130
131	132	133	134	135	136	137	138	139	140
141	142	143	144	145	146	147	148	149	150
151	152	153	154	155	156	157	158	159	160
161	162	163	164	165	166	167	168	169	170
171	172	173	174	175	176	177	178	179	180
181	182	183	184	185	186	187	188	189	190
191	192	193	194	195	196	197	198	199	200

Read down each column of numbers. Can you see a pattern? Each number is **one hundred more** than the number before it.

0	1	2	3	4	5	6	7	8	9
100	101	102	103	104	105	106	107	108	109
200	201	202	203	204	205	206	207	208	209
300	301	302	303	304	305	306	307	308	309
400	401	402	403	404	405	406	407	408	409
500	501	502	503	504	505	506	507	508	509
600	601	602	603	604	605	606	607	608	609
700	701	702	703	704	705	706	707	708	709
800	801	802	803	804	805	806	807	808	809
900	901	902	903	904	905	906	907	908	909

Let's practise

1) Say, then write, the number that is **10 more than**:
   a) 155
   b) 128
   c) 200
   d) 485
   e) 801
   f) 450

2) Say, then write, the number that is **100 more than**:
   a) 807
   b) 501
   c) 613
   d) 340
   e) 898
   f) 404

3) Count forwards in tens.

a)

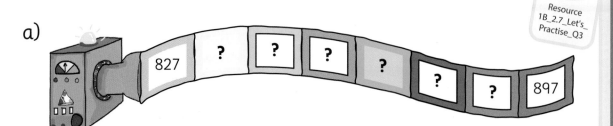

827   ?   ?   ?   ?   ?   ?   897

Resource 1B_2.7_Let's_Practise_Q3

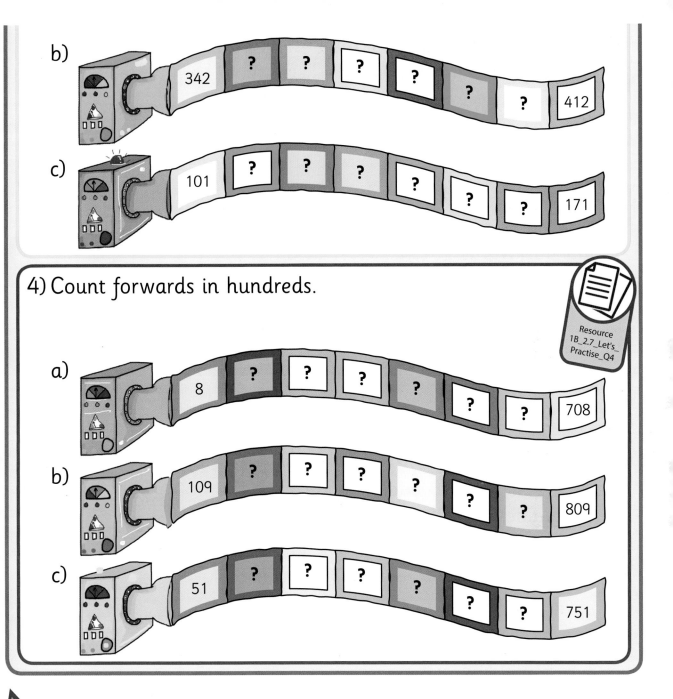

b)

342 | ? | ? | ? | ? | ? | ? | 412

c)

101 | ? | ? | ? | ? | ? | ? | 171

4) Count forwards in hundreds.

Resource 1B_2.7_Let's_Practise_Q4

a)

8 | ? | ? | ? | ? | ? | ? | 708

b)

109 | ? | ? | ? | ? | ? | ? | 809

c)

51 | ? | ? | ? | ? | ? | ? | 751

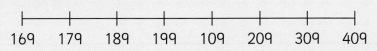

CHALLENGE!

Finlay counted forwards in tens, but he has made some mistakes. Say, then write, what Finlay should have written.

169   179   189   199   109   209   309   409

2.8 Counting backwards in tens and hundreds

We are learning to count backwards in tens and hundreds from a three-digit number.

Before we start

Count backwards in tens. Start at 81.

81

Looking for patterns can help us when counting backwards in tens and hundreds.

Let's learn

Read the numbers in each column from the bottom to the top. Each number is **10 less** than the number before it.

401	402	403	404	405	406	407	408	409	410
411	412	413	414	415	416	417	418	419	420
421	422	423	424	425	426	427	428	429	430
431	432	433	434	435	436	437	438	439	440
441	442	443	444	445	446	447	448	449	450

Read the numbers in each column from the bottom to the top. Each number is **100 less** than the number before it.

1	2	3	4	5	6	7	8	9	10
101	102	103	104	105	106	107	108	109	110
201	202	203	204	205	206	207	208	209	210
301	302	303	304	305	306	307	308	309	310
401	402	403	404	405	406	407	408	409	410

Let's practise

1) Say, then write, the number **10 less than**:
   a) 449
   b) 541
   c) 789
   d) 610
   e) 800
   f) 999
   g) 200
   h) 807
   i) 305

2) Say, then write, the number **100 less than**:
   a) 324
   b) 537
   c) 801
   d) 710
   e) 222
   f) 101
   g) 158
   h) 990
   i) 999

3) Work out the patterns to count backwards down the ladders.

Resource
1B_2.8_Let's_
Practise_Q3

a)

| 462 |
| 452 |
| ? |
| ? |
| ? |
| ? |
| ? |

b)

| 215 |
| 205 |
| ? |
| ? |
| ? |
| ? |
| ? |

c)

| 760 |
| 660 |
| ? |
| ? |
| ? |
| ? |
| ? |

d)

| 504 |
| 404 |
| ? |
| ? |
| ? |
| ? |
| ? |

CHALLENGE!

a) Write down all the different three-digit numbers you can make with these cards.

3 **5** **2**

b) For each number you have made, write the number:
   • 1 less
   • 10 less
   • 100 less

2.9 Counting forwards and backwards in twos and fives

We are learning to count forwards and backwards in twos and fives.

Before we start

Decide whether to count the toys in twos or fives.

a) How many cars?

b) How many marbles?

Being able to spot patterns can help us to count in twos and fives.

Let's learn

Read the numbers on each number line forwards and backwards.

Can you spot the patterns?

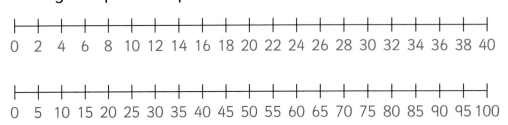

0 2 4 6 8 10 12 14 16 18 20 22 24 26 28 30 32 34 36 38 40

0 5 10 15 20 25 30 35 40 45 50 55 60 65 70 75 80 85 90 95 100

1) a) Isla wants to put two sweets on top of each cake.
How many sweets will she need?

b) Finlay hides five cubes in each cup.
How many cubes does Finlay hide?

c) How much money does Amman have?

d) How much money does Nuria have?

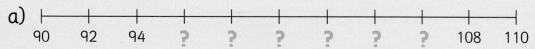

2) The children are counting backwards in fives. What numbers will Finlay, Nuria and Isla say?

 95  ?  ?  ?

CHALLENGE! ....

Find the missing numbers in each sequence.

a)
90   92   94   ?   ?   ?   ?   ?   ?   108   110

b)
?   90   95   ?   ?   ?   115   120   125   ?   ?

2.10 Counting in hundreds, tens and ones

> We are learning to count in hundreds, tens and ones.

Before we start

Isla thinks there are 90 straws altogether.
Explain why she is wrong. How many straws are there?

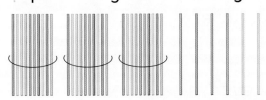

> Grouping objects into tens and hundreds helps make counting easier.

Let's learn

The dots are grouped in hundreds, tens and ones.

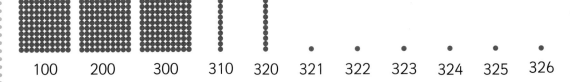

100   200   300   310   320   321   322   323   324   325   326

There are 326 dots altogether.

The blocks are grouped in hundreds, tens and ones.

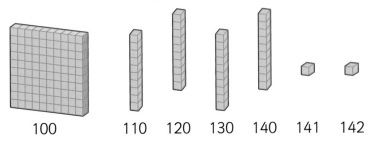

100   110   120   130   140   141   142

There are 142 blocks altogether.

1) Count in hundreds, tens and ones. How many dots altogether?

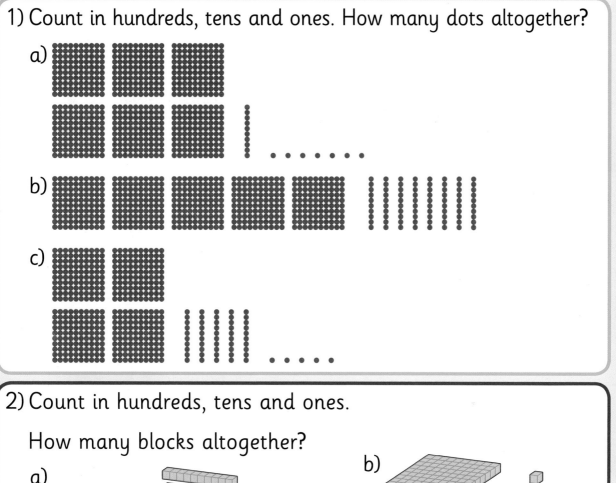

a)

b)

c)

2) Count in hundreds, tens and ones.

How many blocks altogether?

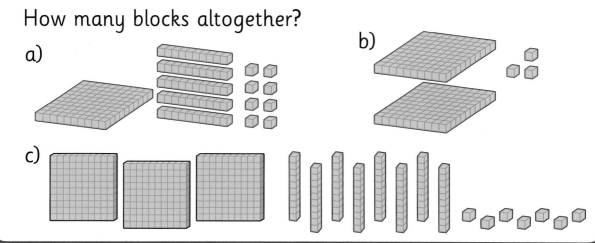

a)

b)

c)

CHALLENGE!

Work with a partner. Make your own models of three-digit numbers with dots or blocks. Challenge your partner to count and write down how many there are altogether. Are they correct?

2.11 Standard place value partitioning

We are learning to partition two-digit numbers.

Before we start

Use pairs of these numbers to make as many two-digit numbers as you can.

 3 **8** **6** **1**

We can **partition** two-digit numbers into tens and ones.

Let's learn

Isla makes the number 34 with ten frames.

30 + 4 = 34

Ten frames

30   4

Finlay makes the same number with straws.

Three bundles of ten and four single ones make 34 altogether

30 + 4 = 34

Amman makes the number 34 using base10 blocks.

30 + 4 = 34

3 tens and 4 ones = 34

Nuria uses place value arrow cards to show that **30 + 4 = 34**.

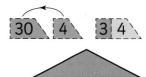

34 is made up of **3 tens** and 4 ones.
We can show this on a place value house.

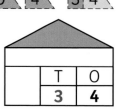

Let's practise

1) Write the number shown by each set of base 10 blocks in three different ways. One has been done for you.

a)  Twenty-seven = 20 + 7 = 27

b)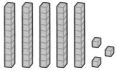

c)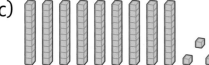

d)

2) Make these two-digit numbers with straws or ten frames then partition them into tens and ones. One has been done for you.

a) 45

   40 + 5

b) 77   c) 32   d) 51   e) 83   f) 68

3) What two-digit numbers can be made with these place value arrow cards? One has been done for you.

a) 1 0 \ and 5 \ makes 15

b) 2 0 \ and 8 \ makes **?**

c) 5 0 \ and 3 \ makes **?**

d) 7 0 \ and 9 \ makes **?**

e) 8 0 \ and 2 \ makes **?**

4) Match each ball to a goal:

 58

40 + 1

41

50 + 8

14

90 + 0

85

10 + 4

90

80 + 5

5) Amman partitions the number 27 in a different way.

27 = 20 + 7

It can also be 10 + 17

 +

Partition each number in a different way.
a) 35 = 30 + 5. It can also be **? + ?**
b) 52 = 50 + 2. It can also be **? + ?**
c) 61 = 60 + 1. It can also be **? + ?**

⭐ CHALLENGE!

How many different numbers can you make using these arrow cards?

2 0	4
3 0	5
8 0	9

2.12 Comparing numbers

We are learning to compare numbers.

Before we start

Finlay thinks he has more marbles than Isla.

Isla thinks Finlay has less than her. Who is correct? Why?

Finlay's marbles   31

Isla's marbles   13

When we compare two numbers we work out which is greater and which is less.

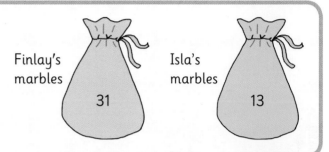

Let's learn

Let's compare the numbers 76 and 67.

Look at the first digit of each number.

76   7 is the tens digit. It means 70.

67   6 is the tens digit. It means 60.

60 is smaller than 70 so **67 is less than 76** and **76 is greater than 67**.

We can use a number line to check.

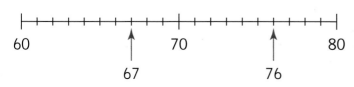

60   70   80
            67   76

Now, let's compare the numbers 342 and 423.

Look at the first digit of each number.

345   3 is the hundreds digit. It means 300.

435   4 is the hundreds digit. It means 400.

400 is greater than 300 so **435 is greater than 345** and **345 is less than 435.**

Number line check:

```
  +++++++++++++++++++++++++++++++++
300            ↑              400        ↑
              345                        435
```

Let's practise

1) The children are playing a game in pairs. The player with the greater number wins. Name the winner in each game. Draw a number line to justify your answer.

a)

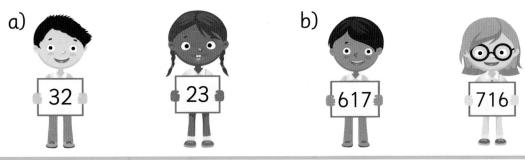

32   23

b)

617   716

2) Choose the correct phrases to complete the number chains.

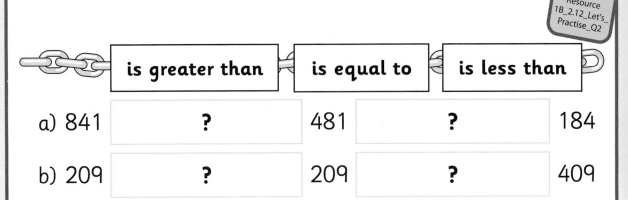

is greater than	is equal to	is less than

a) 841   ?   481   ?   184

b) 209   ?   209   ?   409

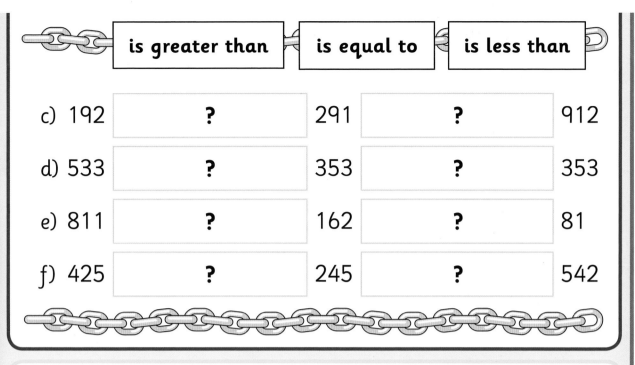

is greater than		is equal to		is less than
c) 192	?	291	?	912
d) 533	?	353	?	353
e) 811	?	162	?	81
f) 425	?	245	?	542

3) Write a number to make each number sentence true.

a) **?** is less than 409.　　b) **?** is greater than 920.

c) **?** is greater than 200.　d) **?** is less than 600.

⭐ **CHALLENGE!** ....

Write sentences about the numbers below.

Use the words '**is greater than**' or '**is less than**'.

How many different sentences can you find? One has been done for you. 695 **is greater than** 659.

956　695　659　965　569　596

2.13 Ordering numbers

We are learning to order numbers.

Before we start

True or false? A number line may help.

a) 62 is greater than twenty-six.

b) Fifty-one is equal to 51.

c) Seventy is less than 17.

d) 245 is less than 425.

Numbers can be ordered from smallest to largest or from largest to smallest.

Let's learn

Let's order these numbers from smallest to largest:

512   **1**25   **3**55   **2**72

Look at the **first digit** of each number and compare them.

125 is the smallest number.   **2**72 is the next smallest, then **3**55. **5**12 is the largest number.

From smallest to largest the numbers are:

125   **2**72   **3**55   **5**12.

When two numbers have the same first digit, we need to look at the second digit to compare them.
For example, 45 has 4 tens and 5 ones. 42 has 4 tens and 2 ones. 45 is larger than 42.

Drawing a number line can help us work out the correct order.

1) These numbers should be in order from smallest to largest but one is in the wrong place. Write down the correct order.

a)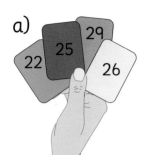

b)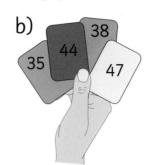

c)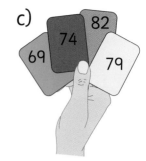

2) Write each set of numbers in your jotter in order from largest to smallest. Use a number line to check.

a)

156   175   149   160   105

b)

230   320   410   190   910

CHALLENGE!

Write at least four three-digit numbers on sticky notes. Give them to a friend to put in the correct order.

2.14 Ordinal numbers

> We are learning to describe the order of things.

Before we start

Write the position of each skittle marked * in both numerals and words.

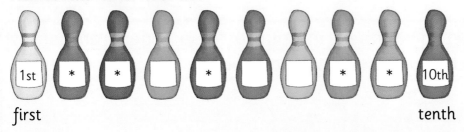

> We can use words and numbers to describe the order of things.

Let's learn

Read the number order words out loud.

1st	2nd	3rd	4th	5th	6th	7th	8th	9th	10th
first	second	third	fourth	fifth	sixth	seventh	eighth	ninth	tenth

11th	12th	13th	14th	15th	16th	17th	18th	19th	20th
eleventh	twelfth	thirteenth	fourteenth	fifteenth	sixteenth	seventeenth	eighteenth	nineteenth	twentieth

Talk about how to write the number order words.

21st	22nd	23rd	24th	25th	26th	27th	28th	29th	30th
?	?	?	?	?	?	?	?	?	?

1) Ted is **1st** in the race.

Bill   Liz   Dave   Jim   Tom   Jen   Jack   Sue   Paul   Mike   Kate   Steve   Kim   Ted

In what position are these drivers?

Write your answers in numerals and in words in your jotter.

a) Dave   b) Sue   c) Liz   d) Jen

e) the driver in front of Kate   f) the driver 2 places behind Jim

g) the last driver in the race

2) You will need Isla's calendar.

July

Sunday	Monday	Tuesday	Wednesday	Thursday	Friday	Saturday
		1	2	3	4	5
6	7 dentist	8	9	10	11	12
13	14	15 lunch with mum	16	17	18	19
20	21	22 cinema	23 doctor	24	25	26
27	28	29	30 Nuria's party	31 picnic		

Resource 1B_2.14_Let's_Practise_Q2

a) Where is Isla going on the twenty-second of July?

b) What is she doing on the thirty-first of July?

c) Who is Isla seeing on the seventh of July?

d) What is she doing on the thirtieth of July?

CHALLENGE!

Sort these ordinal numbers from first to last.
Write each answer in numerals in your jotter.

| thirteenth | ninety-seventh | forty-second | twenty-sixth |

| one hundredth | fourth | sixty-third |

3.1 Compatible numbers: pairs to ten

We are learning to add a string of numbers.

Before we start

The numbers on each bunch of balloons total 10.
Find the missing number on the:

a) red balloon   b) blue balloon   c) green balloon

When we have lots of numbers to add, it helps to look for pairs that make 10.

Let's learn

Here are the number buddies to 10.

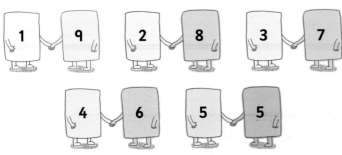

Finlay uses number buddies to help him work out 6 + 3 + 4

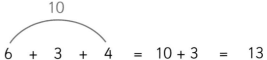

$$6 + 3 + 4 = 10 + 3 = 13$$

1) Use number buddies to help you work out each total.

a) 8 + 6 + 2

b) 9 + 9 + 1

c) 4 + 6 + 6

d) 7 + 4 + 3

e) 5 + 3 + 5

f) 8 + 8 + 2

2) Match up the number buddies and then write the total. What do you notice?

a) (1) + (5) + (5) + (9) = 20

b) 2 + 6 + 8 + 4 = ?

c) 3 + 3 + 7 + 7 = ?

d) 5 + 2 + 5 + 8 = ?

3) Which numbers are missing and how do you know? Write your answers in your jotter.

a) 2 + ? + 3 + 7 = 20

b) 4 + 6 + 1 + ? = 20

c) 5 + 1 + 9 + ? = 20

d) 4 + 8 + ? + 2 = 20

CHALLENGE!

How many different ways can you make 18 using:

a) three numbers

b) four numbers

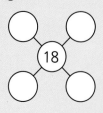

3.2 Doubles and near doubles

> We are learning to use doubles and near doubles to add mentally.

Before we start

Isla is making a necklace. If she **doubles** the number of beads on the string, how many will she have?

Double [**?**] = [**?**]   [**?**] + [**?**] = [**?**]

> We can use double facts to work out near double facts.

Let's learn

These ten frames show double 7.

7 + 7 = 14

Add one more counter. The ten frames now show double 7 plus 1 or 8 + 7

Double 7 plus 1 = 15   8 + 7 = 15

We call 8 + 7 a **near double.**

Let's practise

1) Use double facts to work out each near double.

a) 5 + 6 = double [**?**] plus one = [**?**] + 1 = [**?**]

b) 8 + 9 = double [**?**] plus one = [**?**] + 1 = [**?**]

c) 7 + 8 = double [**?**] plus one = [**?**] + 1 = [**?**]

d) 6 + 7 = double [**?**] plus one = [**?**] + 1 = [**?**]

e) 9 + 10 = double [**?**] plus one = [**?**] + 1 = [**?**]

2) Choose a double fact to help you work out the answers to these additions.

I know that 16 is double 8.
17 is near to 16.
It is double 8 (16) plus 1.

I know that 18 is double 9.
17 is also near to 18.
It is double 9 (18) minus 1.

a) 10 + 9   b) 8 + 7   c) 7 + 6   d) 6 + 5   e) 9 + 8

5 + 5   7 + 7   9 + 9

6 + 6   8 + 8   10 + 10

CHALLENGE!

The numbers around the outside of each triangle add up to 20.
Find the missing double fact then write the number sentence.
The first one has been done for you.

a)
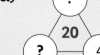
? 20 ? 4

4 + 8 + 8 = 20

b)
? 20 8 ?

c)
2 20 ? ?

d)
0 20 ? ?

3.3 Fact families

> We are learning to find number bonds to 20.

Before we start

Isla writes a fact family for the number buddies 3 and 7.

$3 + 7 = 10$ $7 + 3 = 10$

$10 - 3 = 7$ $10 - 7 = 3$

Write a fact family for each pair of number buddies.

a)

b)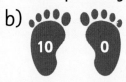

c)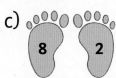

> A fact family is two additions and two subtractions made with the same whole and parts.

Let's learn

Nuria uses cubes to help her write a fact family with the numbers 9 and 6.

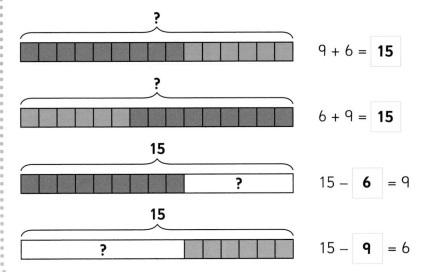

$9 + 6 = \boxed{15}$

$6 + 9 = \boxed{15}$

$15 - \boxed{6} = 9$

$15 - \boxed{9} = 6$

1) Make up three different calculations from the following:

   a) $5 + 8 = 13$ b) $6 + 7 = 13$

   c) $4 + 9 = 13$ d) $3 + 10 = 13$

2) Write three different number sentences to complete each fact family.

   a)   b)   c)   d)

   $8 + 9 = 17$ $7 + 5 = 12$ $9 + 7 = 16$ $8 + 6 = 14$

3) Finlay writes a fact family for this **bar model**.

whole

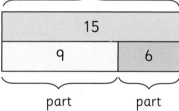

| 15 | |
| 9 | 6 |

part   part

$9 + 6 = 15$ $6 + 9 = 15$

$15 - 9 = 6$ $15 - 6 = 9$

Write a fact family for each bar model.

a)
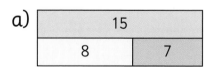

| 15 | |
| 8 | 7 |

b)

| 17 | |
| 9 | 8 |

c)

| 11 | |
| 2 | 9 |

d)

| 12 | |
| 4 | 8 |

4) Write down as many number pairs that total 15 as you can. Now swap your work with a partner. Did they think of any number pairs that you didn't get? Work together to write a subtraction fact for each addition fact.

5) Find the missing numbers: Draw a bar model to help.

a) ⭐ **?** + 9 = 15

b) ⭐ **?** − 13 = 3

c) 18 − ⭐ **?** = 9

d) 7 + ⭐ **?** = 18

e) ⭐ **?** − 7 = 12

f) ⭐ **?** + 8 = 17

6) Use these digits to make up six missing number statements.

| 8 | 4 | 9 | 16 | 17 |

For example: 16 − **?** = 4

Ask a partner to find each missing number.

⭐ **CHALLENGE!** ....

Isla was asked to solve this missing number sentence:

20 − **?** = **?**

She was given the clue that the solution was an even number less than 10.

Write down all the possible number sentences Isla could have written.

3.4 Complements to the next multiple of 10

We are learning to count on and back to a multiple of ten.

Before we start

Partition each amount into tens and ones. One has been done for you.

a)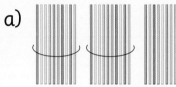

20 + 8 = 28

b)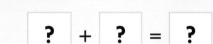

? + ? = ?

c)

? + ? = ?

d)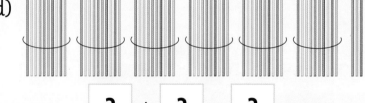

? + ? = ?

e)

? + ? = ?

Number bonds can help us to count on and back.

Let's learn

The tens numbers are sometimes called **multiples** of ten.

| 10 | 20 | 30 | 40 | 50 | 60 | 70 | 80 | 90 |

Amman **counts on** from 74 to the next multiple of 10.

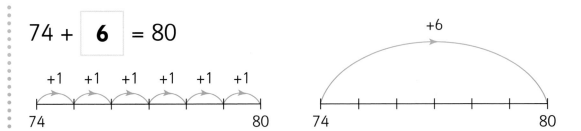

74 + **6** = 80

Finlay **counts back** from 38 to the next multiple of ten.

38 − **8** = 30

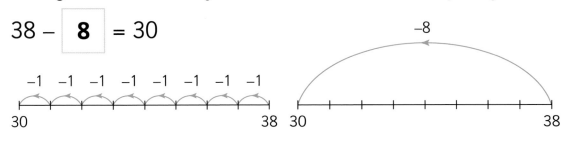

1) Add a number from the red truck to each number from the yellow truck to make multiples of ten. Write a number sentence for each addition. For example: **57 + 3 = 60**.

57   13   45   42   78
   26   31   86   66   77
53   68   51   82   65

2   4   7   9
3   5   8

2) Choose a number from the first truck and subtract a number from the second truck to make a multiple of 10. Write a number sentence for each subtraction. For example:
47 − 7 = 40.

⭐ **CHALLENGE!** ....

Work with a partner. Write a multiple of ten in each box to make the number sentences true. How many different ways can you find?

a) 5 + [?] + [?] = 75   b) 84 − [?] − [?] = 4

3.5 Adding 0, 1 and 10

> We are learning to add 0, 1 and 10 to a number.

Before we start

Find the missing numbers.

a) 50 3 $50 + 3 =$ [?]

b) 30 [ ] $30 +$ [?] $= 35$

c) [ ] 9   [?] $+ 9 = 79$

d) 90 5   [?] $+$ [?] $=$ [?]

> We can count forwards in tens and ones to help us with addition.

Let's learn

Finlay and Nuria are working out the answer to **10 + 62**

Finlay draws a number line, then counts in tens and ones. He uses a bead string to check his answer.

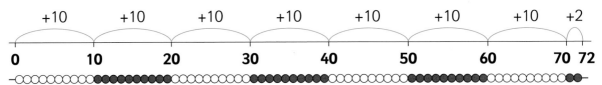

> If I add one more bead there will be 73 altogther.
>
> $72 + 1 = 73$

Nuria thinks 62 + 10 = ? She uses a 100 square to find the answer.

When we add zero the starting number doesn't change.

72 + 0 = 72

1	2	3	4	5	6	7	8	9	10
11	12	13	14	15	16	17	18	19	20
21	22	23	24	25	26	27	28	29	30
31	32	33	34	35	36	37	38	39	40
41	42	43	44	45	46	47	48	49	50
51	52	53	54	55	56	57	58	59	60
61	62	63	64	65	66	67	68	69	70
71	72	73	74	75	76	77	78	79	80
81	82	83	84	85	86	87	88	89	90
91	92	93	94	95	96	97	98	99	100

Let's practise

1) a) 85 + 0 = ?   85 + 1 = ?   85 + 10 = ?

b) 27 + 0 = ?   27 + 1 = ?   27 + 10 = ?

c) ? + 0 = 39   ? + 1 = 40   ? + 10 = 49

d) ? + 0 = 76   ? + 1 = 77   ? + 10 = 86

e) 44 + ? = 44   44 + ? = 45   44 + ? = 54

f) 91 + ? = 91   91 + ? = 92   91 + ? = 101

2) Use your answers to question 1 to help you work out:

    a) 85 + 11   b) 27 + 11   c) 39 + 11

    d) 75 + 11   e) 44 + 11   f) 91 + 11

3) Find the missing numbers for each function machine.

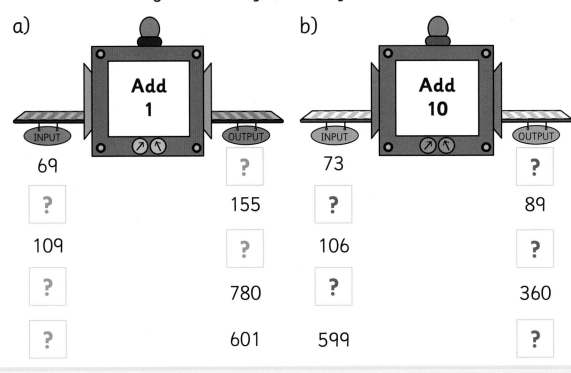

a)

INPUT	OUTPUT
69	?
?	155
109	?
?	780
?	601

b) Add 10

INPUT	OUTPUT
73	?
?	89
106	?
?	360
599	?

CHALLENGE!

1) For each example, write the number hiding behind the:
- green splat
- red splat
- blue splat

a)

52	🔆	54
🔆	63	🔆

b)

17	18	🔆
27	🔆	🔆

c)

28	29	🔆
🔆	39	🔆

2) For each example, what must be added to the purple number to make the blue splat number?

3.6 Subtracting 0, 1 and 10

> We are learning to subtract 0, 1 or 10 from a number.

Before we start

1) Here is part of a 100 square. What number is hiding behind the:

    a) red splat   b) purple splat

    c) blue splat   d) green splat

    e) yellow splat

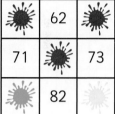

2) What do you notice about the numbers in:

    a) each row   b) each column

> We can count back in tens or ones to help us with subtraction.

Let's learn

Isla and Amman are working out the answer to **62 – 10**.

Isla draws a number line then counts back in tens and ones.
She uses a bead string to check her answer.
Talk about Isla's strategies.

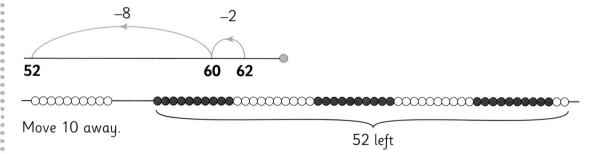

Move 10 away.

52 left

Amman finds 62 – 10 on a 100 square.

1	2	3	4	5	6	7	8	9	10
11	12	13	14	15	16	17	18	19	20
21	22	23	24	25	26	27	28	29	30
31	32	33	34	35	36	37	38	39	40
41	42	43	44	45	46	47	48	49	50
51	52	53	54	55	56	57	58	59	60
61	62	63	64	65	66	67	68	69	70
71	72	73	74	75	76	77	78	79	80
81	82	83	84	85	86	87	88	89	90
91	92	93	94	95	96	97	98	99	100

$62 - 10 = 52$

I can also take away 1 by jumping to the number before 62.

$62 - 1 = 61$

When we subtract zero the starting number stays the same so $62 - 0 = 62$.

Let's practise

1) Copy and complete the patterns in your jotter.

a) $87 - 10 =$ **?**

$77 - 10 =$ **?**

$67 - 10 =$ **?**

$57 - 10 =$ **?**

$47 - 10 =$ **?**

b) $63 - 1 =$ **?**

$53 - 1 =$ **?**

$43 - 1 =$ **?**

$33 - 1 =$ **?**

$23 - 1 =$ **?**

c) $41 - 0 =$ **?**

$31 - 0 =$ **?**

$21 - 0 =$ **?**

$11 - 0 =$ **?**

$1 - 0 =$ **?**

2) Find the missing numbers for each function machine.

a)

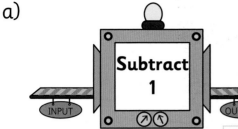

INPUT	OUTPUT
54	**?**
?	70
303	**?**
?	900
?	110

b)

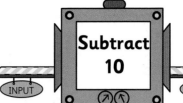

INPUT	OUTPUT
29	**?**
?	68
465	**?**
?	571
?	700

3) a) 92 – 1 = **?**   b) **?** = 69 – 0   c) 75 – **?** = 65

d) 26 – **?** = 25   e) **?** – 0 = 41   f) **?** – 10 = 38

CHALLENGE! ....

Isla starts with **51**. She **subtracts 10** to reach **41**. Then she **subtracts 1** to reach **40**.

a) How much did Isla subtract altogether? 51 – **?** = 40

b) Here are five number cards

87	55	62	31	90

Subtract 11 from each number.

3.7 Adding and subtracting single and double-digit numbers

> We are learning to use number patterns and known facts to add and subtract.

Before we start

Check Finlay's work. For each wrong answer, write what Finlay should have written in your jotter.

a) 6 + 3 = 9   b) 8 – 4 = 4   c) 10 + 2 = 102

d) 90 + 5 = 59   e) 11 – 6 = 7   f) 9 – 7 = 2

g) 6 + 20 = 26   h) 9 + 8 = 18   i) 10 – 5 = 15

> We can use number patterns, and the addition and subtraction facts we know, to help us add and subtract other numbers.

Let's learn

Nuria uses the number fact 3 + 4 = 7 to work out 13 + 4 = 17.

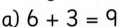

3 + 4 = 7

13 + 4 = 17

23 + 4 = 27

33 + 4 = 37

> 13 is ten more than 3, so the answer to 13 + 4 must be ten more that the answer to 3 + 4.

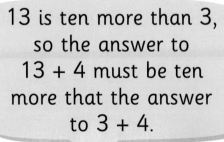

Nuria's strategy works for subtraction too!

Can you see a pattern? What would the next subtraction be?

$9 - 8 = 1$ $19 - 8 = 11$ $29 - 8 = 21$ $39 - 8 = 31$

$49 - 8 = 41$ $59 - 8 = 51$ $69 - 8 = 61$ $79 - 8 = 71$

Let's practise

1) Copy and complete the patterns in your jotter, then write the next three number sentences for each pattern.

a) $4 + 5 =$ **?**

   $14 + 5 =$ **?**

   $24 + 5 =$ **?**

b) $2 + 6 =$ **?**

   $12 + 6 =$ **?**

   $22 + 6 =$ **?**

c) $8 - 5 =$ **?**

   $18 - 5 =$ **?**

   $28 - 5 =$ **?**

d) $7 - 3 =$ **?**

   $17 - 3 =$ **?**

   $27 - 3 =$ **?**

2) a) $5 + 1 =$ **?**

   $45 + 1 =$ **?**

   $75 + 1 =$ **?**

b) $3 + 6 =$ **?**

   $63 + 6 =$ **?**

   $93 + 6 =$ **?**

c) $8 - 4 =$ **?**

   $38 - 4 =$ **?**

   $88 - 4 =$ **?**

d) $10 - 5 =$ **?**

   $20 - 5 =$ **?**

   $50 - 5 =$ **?**

3) Complete these calculations. Then write down in your jotter the known fact that you used to help you.

For example:

6 + 42 = **48**   6 + 2 = 8

a) 3 + 53 = **?**   b) 76 + **?** = 79   c) 22 + 5 = **?**

d) 95 – 2 = **?**   e) 19 – 5 = **?**   f) 46 – **?** = 40

g) **?** + 34 = 38   h) 27 – 6 = **?**   i) 69 – **?** = 66

j) 82 + **?** = 88   k) **?** + 41 = 44   l) 38 – **?** = 35

⭐ CHALLENGE!

Your starting number is 45.

What is your final number?

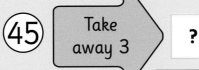

45 → Take away 3 → **?**

Add 5

? → Subtract 2 → **?** → Plus 4 → **?**

Now, choose a different two-digit number and play the game again.

3.8 Adding and subtracting multiples of 10 to and from two-digit numbers

> We are learning to add and subtract multiples of ten.

Before we start

1) Match pairs of arrow cards to make the numbers below. One has been done for you.

   a) 56 = 50 + 6

   b) 27 = **?** + **?**

   c) 63 = **?** + **?**

   d) 85 = **?** + **?**

   20   5   60

   50   6   3

   8   80   7

2) Which arrow card does not have a partner? Draw an arrow card that could be joined with it to make a number that is more than 30 but less than 50.

> We can use known facts and place value to add and subtract multiples of ten.

Let's learn

Isla counts out 48 straws and 30 straws.

She counts on from 48 to find how many straws altogether.

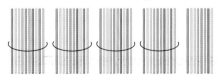 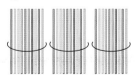

Isla draws a number line to show how she worked out her answer.

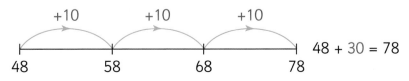

+10   +10   +10

48   58   68   78   48 + 30 = 78

Nuria makes the number 62 with tens and ones strips. She covers up **2 tens** (20).

Nuria draws a number line to show what she did.

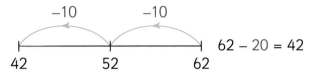

62 − 20 = 42

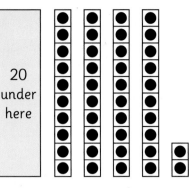

20 under here

Let's practise

1) Use known facts to work out:
   a) 50 + 20
   b) 40 + 60
   c) 10 + 20
   d) 80 − 30
   e) 100 − 20
   f) 70 − 40

2) Count on or back in tens to work out each answer. An empty number line may help.
   a) 90 + 30
   b) 80 + 60
   c) 50 + 70
   d) 110 − 20
   e) 130 − 90
   f) 150 − 80

3) Complete these addition puzzles.

a)
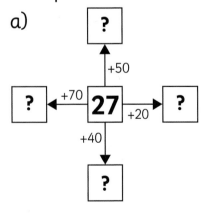

b)
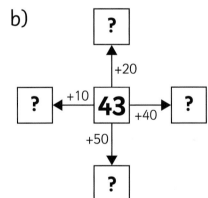

Number lines

4) Complete these subtraction puzzles.

a)

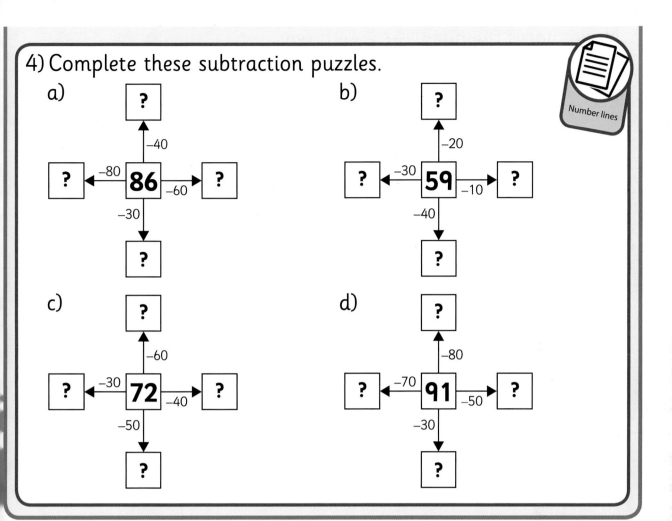

?

−40

? ←−80− **86** −60→ ?

−30

?

b)

?

−20

? ←−30− **59** −10→ ?

−40

?

c)

?

−60

? ←−30− **72** −40→ ?

−50

?

d)

?

−80

? ←−70− **91** −50→ ?

−30

?

CHALLENGE!

Both number lines show that 97 + 30 = 127.

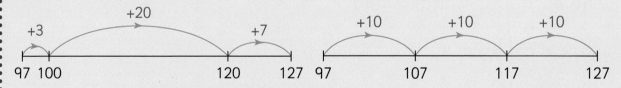

+20

+3   +7

+10   +10   +10

97 100   120   127   97   107   117   127

Draw number lines to find the answers to these calculations.
Compare number lines with a partner.
Did you both work them out in the same way?

a) 385 + 50   b) 232 − 40   c) 718 − 60

3.9 Partitioning through 10

We are learning to use an empty number line to add and subtract.

Before we start

Which is the odd one out? Justify your answer.

a)  55 5

b)  41 9

c)  36 4

d)  72 8

e)  27 2

f)  83 7

g)  64 6

An empty number line can help us with addition and subtraction.

Let's learn

Isla draws an empty number line to help her calculate 78 + 9

She partitions 9 into 2 and 7.

78 **+ 2** → 80 **+ 7** → 87 so

78 **+ 9** = 87

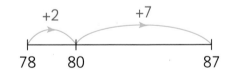

Nuria draws an empty number line to help her calculate 63 – 7.

She partitions 7 into 3 and 4.

63 **– 3** → 60 **– 4** → 56

63 **– 7** = 56

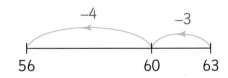

Let's practise

1) Draw an empty number line for each addition then solve it.

  Number lines

a) 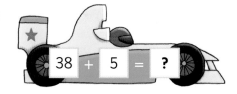 38 + 5 = ?

b)  45 + 7 = ?

c)  26 + 5 = ?

d)  48 + 6 = ?

e)  36 + 5 = ?

f)  65 + 8 = ?

2) Answer these additions. Show your method on empty number lines.

a) 57 + 6 b) 68 + 7 c) 49 + 5

d) 74 + 7 e) 89 + 7 f) 95 + 6

3) Draw an empty number line for each subtraction then solve it.

a)  23 − 5 = ?

b)  55 − 6 = ?

c)  33 − 5 = ?

d)  52 − 6 = ?

e)  44 − 5 = ?

f)  75 − 8 = ?

4) Answer these subtractions. Show your method on empty number lines.

a) 94 − 7

b) 81 − 4

c) 63 − 8

d) 55 − 6

e) 22 − 9

f) 42 − 5

5) Draw a number line to solve each calculation.

a) 78 + **?** = 86

b) 54 + **?** = 63

c) 91 = 83 + **?**

d) 76 = 83 − **?**

e) 46 = 54 − **?**

f) 72 − **?** = 66

g) 47 = 39 + **?**

h) 31 − **?** = 24

i) 36 + **?** = 45

j) 93 − **?** = 85

⭐ CHALLENGE! ....

Work with a partner. Using these number cards, how many subtractions can you find that make multiples of 10?

You can use the cards **as many times as you like** in your subtractions.

3.10 Adding two-digit numbers

We are learning to add two-digit numbers.

Before we start

Jump along the number line in tens. Say and write down the missing numbers in your jotter.

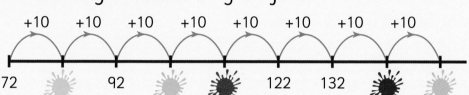

+10   +10   +10   +10   +10   +10   +10   +10

72   92   122   132

Partitioning can help us to add two-digit numbers.

Let's learn

Amman and Finlay jump along an empty number line to work out **28 + 36**. They keep the first number whole and partition the second number.

Amman's number line:

+10   +10   +10
                              +2 +4

28   38   48   58 60 64

$$28 + 36 = 64$$

Finlay's number line:

        +10   +10   +10
+2   +4

28 30   40   50   60 64

$$28 + 36 = 64$$

Talk about each boy's method. Which do you prefer?

Is there another way?

1) Draw an empty number line to find each total.

    a) 14 + 57   b) 36 + 16

    c) 28 + 17   d) 43 + 18

    e) 65 + 26   f) 27 + 45

    g) 72 + 39   h) 88 + 34

2) Copy each number line. Fill in the missing numbers, then write the number sentence.

    a)

    b)

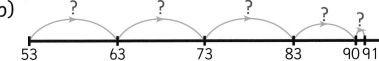

3) Partition both numbers to work out:

    a) 35 + 14   b) 23 + 55   c) 51 + 38   d) 21 + 44

    e) 15 + 71   f) 67 + 22   g) 12 + 65   h) 33 + 34

    i) 18 + 41   j) 47 + 35   k) 18 + 57   l) 29 + 39

Partitioning is a useful strategy for adding two-digit numbers.

Let's work out **25 + 13** by partitioning both numbers.

25 can be partitioned into **20 + 5**
13 can be partitioned into **10 + 3**

20 + 10 = **30** and
5 + 3 = **8**
30 and 8 makes **38**
So **25 + 13 = 38**

Show on an empty number line how you would work out the answers to these additions. Compare your method with a partner.

a) ? + 27 = 42

b) 18 + ? = 31

c) ? + 41 = 60

d) ? + 55 = 84

e) 33 + ? = 56

f) ? + 46 = 92

3.11 Subtracting two-digit numbers

> We are learning to subtract two-digit numbers.

Before we start

True or false?

a) 70 – 40 = 30   b) 60 – 30 = 30   c) 60 – 20 = 80

d) 90 – 50 = 30   e) 80 – 60 = 60   f) 100 – 70 = 30

Turn each false number sentence into a true number sentence by changing one number **or** a symbol.

> Partitioning can help us to subtract two-digit numbers.

Let's learn

Isla and Nuria jump along an empty number line to work out **63 – 28**. They keep the first number whole and partition the second number.

Isla's number line:

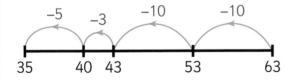

63 – 28 = 35

Nuria's number line:

63 – 28 = 35

Talk about each girl's method. Which do you prefer? Is there another way?

1) Draw an empty number line to find each difference.

a)

b)

c)

d)

e)

f)

g)

h)

2) Copy each number line. Fill in the missing numbers then write the number sentence.

a)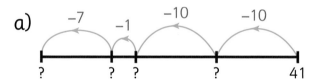

b)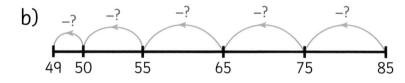

3) Partition both numbers to work out:

a) 38 – 13   b) 67 – 23   c) 79 – 32
d) 26 – 15   e) 44 – 21   f) 57 – 16
g) 86 – 24   h) 99 – 44   i) 97 – 53

> Partitioning is a useful strategy for subtracting two-digit numbers.
>
> Let's work out **46 – 21** by partitioning both numbers.

> 46 can be partitioned into **40 + 6**
>
> 21 can be partitioned into **20 + 1**
>
> 40 – 20 = **20** and
> 6 – 1 = **5** → 20 and
> 5 makes **25** so
> **46 – 21 = 25**

CHALLENGE!

Explain to Finlay why he is wrong.

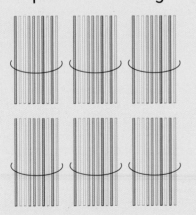

> **65 – 36**
>
> I partitioned both numbers and got the answer **31**.

Draw an empty number line to find the correct answer to 65 – 36.

3.12 Inverse relationships

We are learning to choose between counting on and counting back to solve missing number problems.

Before we start

Follow the arrows to work out the missing numbers.
How will you know if you are correct?

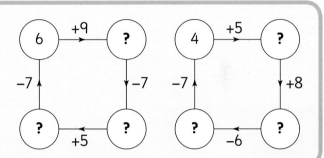

We can check the answer to a subtraction by adding.

Let's learn

Amman uses base ten blocks to prove that **67 – 34 = 33**

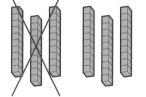

60 – 30 = 30   7 – 4 = 3   67 – 34 = 33

He checks his answer by adding.

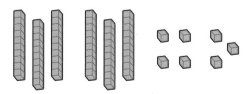

30 + 30 = 60   4 + 3 = 7   34 + 33 = 67

Isla finds the answer to 67 − 34 by counting back on a number line.

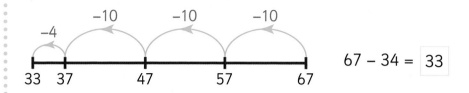

67 − 34 = 33

She checks her answer by counting on.

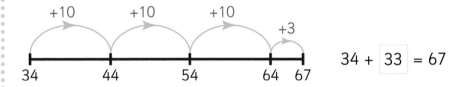

34 + 33 = 67

1) Draw an empty number line to solve each subtraction.
Check each answer by counting on.

a) 48 − 24 = ?   24 + ? = 48

b) 69 − 53 = ?   53 + ? = 69

c) 75 − 31 = ?   31 + ? = 75

d) 97 − 79 = ?   79 + ? = 97

e) 83 − 69 = ?   69 + ? = 83

f) 65 − 38 = ?   38 + ? = 65

g) 52 − 27 = ?   27 + ? = 52

h) 90 − 46 = ?   46 + ? = 90

2) Write a subtraction and an addition for each number line.

a)

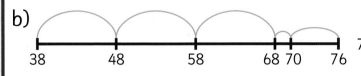

17 20   30 33

33 − ⬚ **?** = 17   17 + ⬚ **?** = 33

b)

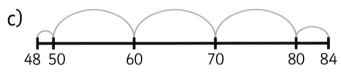

38   48   58   68 70   76

76 − ⬚ **?** = 38   38 + ⬚ **?** = 76

c)

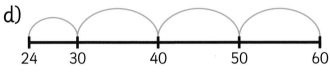

48 50   60   70   80 84

84 − ⬚ **?** = 48   48 + ⬚ **?** = 84

d)

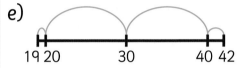

24   30   40   50   60

60 − ⬚ **?** = 24   24 + ⬚ **?** = 60

e)

19 20   30   40 42

42 − ⬚ **?** = 19   19 + ⬚ **?** = 42

f)

57 60   70   80 81

81 − ⬚ **?** = 57   57 + ⬚ **?** = 81

3) Find the missing number by counting on or counting back. Write the complete number sentence.

a) 50 = 35 + ⬚ **?**   b) 61 = ⬚ **?** + 25   c) ⬚ **?** + 18 = 41

d) 90 − ⬚ **?** = 11   e) 32 − ⬚ **?** = 16   f) 36 = 72 − ⬚ **?**

CHALLENGE! ....

Make some addition and subtraction number sentences using just these numbers.

| **42** | **29** | **71** |

3.13 Strategies for solving number puzzles

We are learning to solve number puzzles.

Before we start

Complete each puzzle both clockwise and anti-clockwise.
Which way was easier? Why?

a)

```
 9  ──  −2  ──  ?
 │               │
−9              +3
 │               │
 ?  ──  +8  ──  ?
```

b)

```
 7  ──  −4  ──  ?
 │               │
+3              +9
 │               │
 ?  ──  −8  ──  ?
```

There are different strategies that we can use to help us solve number puzzles.

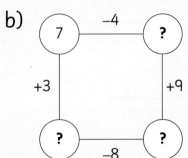

Let's learn

Think about **5 + 4 + 6 =** [**?**]

Here are some different strategies to solve this problem.

Make 10

5 + 4 + 6   6 + 4 = 10   **10 + 5 = 15**

Use doubles or near doubles:

5 + 4 + 6   4 + 4 = 8 so 5 + 4 = 9   9 **+ 6 = 15**

Partition one or more numbers:

5 + 4 + 6   **5 + 4 = 9**   6 = 1 + 5   **9 + 1 + 5 = 15**

Use a number line:

Or you could draw your own strategy on an empty number line.

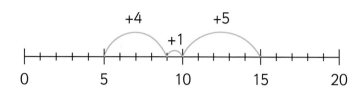

1) Write a number in each circle so that each line has the same total.

a)
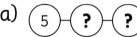

5 – ? – ?
4   5
6 – ? – 7

b)
? – 4 – 7
?   2
? – 5 – 3

c)
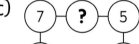

7 – ? – 5
?   ?
8 – 3 – 9

d)
9 – 6 – 3
?   ?
? – 6 – 8

CHALLENGE!

Work with a partner.

Choose a number between 11 and 20.
Challenge your partner to write down as many facts about your number as they can.
Are they all correct? Can you think of others?

3.14 Representing and solving word problems (1)

> We are learning to think about the same word problem in different ways.

Before we start

a) Find the difference between 11 and 9.

11	
9	?

$11 - 9 = \boxed{?}$ $9 + \boxed{?} = 11$

b) Now write a word problem to fit the completed bar model.

> A think board can help us to **represent** and solve word problems.

Let's learn

Talk about Isla's Think Board.

There are 14 beanbags and 8 hoops. If Isla puts one beanbag in each hoop, how many hoops won't have a beanbag? 	**Picture** 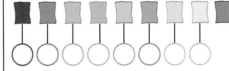
Bar model | 14	| |---|---| | 8 | ? |	**Number sentence and answer** $8 + \boxed{6} = 14$

Draw a think board for each problem, then solve it.

1) There are 18 cups and 11 saucers. How many cups won't have a saucer?

2) There are 15 children in the gym. Miss Burke gives each child a ball and has five balls left over. How many balls are there?

3) Finlay has 20 toy cars and some vans. He has four fewer vans than cars. How many vans does Finlay have?

4) Isla has 19 straws and some bottles. She puts one straw in each bottle and has three straws left. How many bottles does Isla have?

5) There are 13 boys and some girls in Amman's class. There are 4 more girls than boys. How many girls are in the class?

6) Nuria has 25 red beads and 18 blue beads. How many more red beads does Nuria have?

⭐ CHALLENGE! ....

a) Write two addition number sentences and two subtraction number sentences to fit the bar model.

17	
9	8

b) Write a word problem for each number sentence you have made. Ask a partner to solve them.

3.15 Representing and solving word problems (2)

We are learning to think about the same word problem in different ways.

Before we start

Isla had some strawberries in her lunch box. She ate seven strawberries and had five strawberries left. How many strawberries did Isla start with?

A think board can help us to **represent** and solve word problems.

Let's learn

Talk about Finlay's think board.

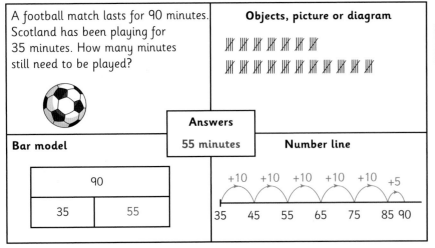

A football match lasts for 90 minutes. Scotland has been playing for 35 minutes. How many minutes still need to be played?	**Objects, picture or diagram**
Bar model	**Answers** 55 minutes	**Number line**

$35 + \boxed{?} = 90$

Draw a think board for each problem in your jotter, then solve it.

1) Isla is saving up for a new bike. She has £58 in her piggy bank. Her gran gives her £37. How much money does Isla have now?

2) Nuria had 33 sweets. She gave some to Isla. Now she has 18. How many sweets did Nuria give to Isla?

3) By lunchtime, Ben the Baker had sold 67 of the cakes he made in the morning. There are only 13 cakes left. How many cakes had Ben baked?

4) Jim has 74 cows. 35 cows are outside. The rest are in the barn. How many cows are in the barn?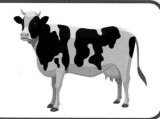

5) Amman has been playing his new computer game. On Saturday he scored 37 points. On Sunday he scored 69 points. How many more points did Amman score on Sunday?

6) Finlay counts 54 cars on his way to school. 26 of them are white. How many cars are not white?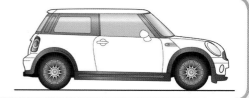

CHALLENGE!

Isla's new jacket cost £28 and her hoody cost £13. If she pays with three £20 notes how much change will she get?

4.1 Multiplying by skip counting

> We are learning to skip count in threes to solve multiplication problems.

Before we start

Isla, Nuria and Amman have £10 each. How much money do they have altogether?

> We can use skip counting to help us multiply by 3.

Let's learn

When we skip count in threes, we skip over numbers to just count the multiples of 3.

We can use number lines to help us:

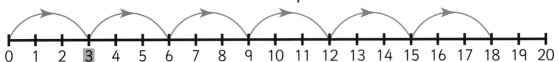

We can also use bead strings:

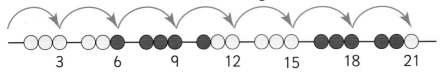

If we are working out 6×3, we can skip count in threes to solve the problem:

3 ... 6 ... 9 ... 12 ... 15 ... 18

$6 \times 3 = 18$

1) a) Use a 100 square. Skip count in threes and put a counter on each multiple of 3 and stop when you get to 30. Can you see a pattern?

b) Which of these numbers do you think will have a counter on them if you continue?

1	2	3	4	5	6	7	8	9	10
11	12	13	14	15	16	17	18	19	20
21	22	23	24	25	26	27	28	29	30
31	32	33	34	35	36	37	38	39	40
41	42	43	44	45	46	47	48	49	50
51	52	53	54	55	56	57	58	59	60
61	62	63	64	65	66	67	68	69	70
71	72	73	74	75	76	77	78	79	80
81	82	83	84	85	86	87	88	89	90
91	92	93	94	95	96	97	98	99	100

100 square

 63   45   92   40 51   75   80   99

Continue putting the counters on the hundred square to see if you are right!

2) Skip count in threes to work out how many cars are in the carpark. You can use a number line or bead string to help you.

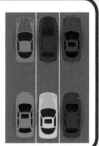

a) Three rows of three cars   b) Five rows of three cars
c) Six rows of three cars   d) Eight rows of three cars

3) Finlay has made biscuits. He has seven bags and he puts three biscuits in each bag. How many biscuits has he got altogether?

⭐ CHALLENGE! ....

Try using skip counting to work out these problems:
a) 9×3 b) 11×3 c) 14×3
d) How many threes are in 36?

4.2 Understanding multiplication as repeated addition

We are learning to recognise and represent multiplication problems as repeated addition.

Before we start

Isla has three pots of pencils.
Each of the pots has five pencils.
Can you make an array to show this problem?

We are learning that multiplication means adding the same number repeatedly.

Let's learn

Isla, Finlay and Nuria have five sweets each.
How many sweets do they have altogether?

We can make an array to help us work it out:

There are three rows of five. We could write the problem like this:

5 + 5 + 5 = 15

We have added the same number three times.

Or we can write the problem like this:

3 × 5 = 15

Adding the same number three times gives the same answer as multiplying by 3.

1) Write each problem as an addition sentence.
Then find the total.

a) There are four shirts. Each shirt has five stripes.
How many stripes in total?

> 5 + 5 + 5 + 5 = 20

b) There are three ladybirds. Each ladybird has
six spots. How many spots in total?

c) There are six houses. Each house has
four windows. How many windows in total?

d) There are seven ships. Each ship has a crew of
seven sailors. How many sailors in total?

e) There are six books. Each book has
12 pages. How many pages in total?

2) The children have been planting some
vegetables. Make or draw an array to show
what they planted and use this to work out
how many altogether. The first one has been
done for you.

2 × 4 = 8

a) Two rows of four carrots   b) Three rows of six potatoes
c) Four rows of three cabbages d) Three rows of five leeks

CHALLENGE!

1) Use squared paper. Colour in the correct number of
squares in the correct number of rows to help you
solve this sum:

9 × 6 =

Squared paper

4.3 Arrays

> We are learning to record arrays as multiplication sentences.

Before we start

Nuria is skip counting in tens to find out how many cubes. She gets the answer 50. Is she right?

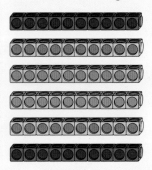

> We can write a multiplication sentence for an array and use skip counting to solve the problem.

Let's learn

There are four rows of five counters in this array.

We can use the multiplication symbol × to write a number sentence like this:

4 × 5

Then we can skip count to work out the answer:

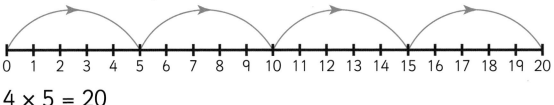

4 × 5 = 20

1)

Write a multiplication sentence for this array.
Skip count to work out the answer.

2) Write a multiplication sentence for this array.
Skip count to work out the answer.

3) Write a multiplication sentence for this array.
Skip count to work out the answer.

4) Finlay has nine pairs of socks. How many altogether?

Make an array to show this problem then write it as a
multiplication sentence. Skip count to work out the answer.

CHALLENGE!

Amman has 24 marbles. Can you find how many different
arrays he could make 24 using these marbles?

Write a multiplication sentence for each array you make.

Use cubes or counters to help you.

4.4 Repeated addition

We are learning to use repeated addition to solve multiplication problems.

Before we start

Finlay has six packets of stickers.
There are five stickers in each packet.

a) Make an array to show the problem and write this as a number sentence.

b) How many stickers does he have altogether?

We can multiply using addition facts that we know.

Let's learn

Isla is working out this multiplication problem:

$3 \times 5 = ?$

I know that double 5 is 10. Then I can add 5 more to get the answer. $10 + 5 = 15$

$3 \times 5 = 15$

1)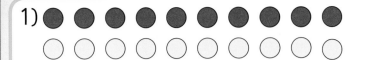

a) Use counters to make a copy of this array.
Write a multiplication sentence for this array.
How many altogether?

b) Add another row of 10 to your array. Write a multiplication sentence for this new array. How many now?

c) Add one more row of 10. How many now? Write a multiplication sentence to show your answer.

2) Finlay has four pots of pencils. Each pot has five pencils.

What adding facts do you know that could help you with this problem? Could you use them to work out how many pencils there are altogether? You could make an array to help you.

3) Isla plants some bulbs in the garden. Work out how many bulbs she has planted altogether.

a) Three rows of two   b) Four rows of 10

c) Five rows of five

★ CHALLENGE! ....

a) Work out the answer to this problem: 10×5

b) Can you use your answer to help you find the answer to these problems?

i) 9×5 ii) 11×5 iii) 12×5 iv) 20×5

4.5 Multiplying using known facts

> We are learning to solve multiplication problems using the facts we already know.

Before we start

Isla has four £1 coins. Nuria has double that number.

How many £1 coins does Nuria have?

> We can use what we know about doubles and tens to help us work out multiplication problems.

Let's learn

Finlay is working out 2 × 4. He makes an array.

He notices that 2 × 4 is the same as doubling 4.

Double 4 = 8, so 2 × 4 = 8.

Isla is working out 5 × 10. She uses an abacus to show the problem.

Isla knows that there are 5 groups of 10 in 50, so she uses this fact to answer the problem.

5 groups of 10 = 50, so 5 × 10 = 50

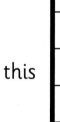

1) Look at these arrays and write a double fact and a multiplication sentence for each one in your jotter. The first one has been done for you.

a) ● ● ●
   ○ ○ ○

| double 3 = 6 |

b) ○ ○ ○ ○ ○
   ● ● ● ● ●

| 2 × 3 = 6 |

c) ● ●
   ○ ○

d) ● ● ● ● ● ● ●
   ● ● ● ● ● ● ●

2) Amman has made brownies.

He puts 10 in each box.
How many?

a) 4 boxes of 10
c) 6 boxes of 10

b) 2 boxes of 10
d) 9 boxes of 10

CHALLENGE!

Work with a partner. Think of a double fact you know and tell your partner just the answer. Then ask them to give you the double fact to match your answer.

What double gives the answer 10?

Double 5 makes 10.

Can they give the multiplication sentence for the double fact too?

Find the double facts for all these numbers: | 2 × 5 = 10 |

2, 4, 6, 8, 10, 12, 14, 16, 18, 20, 30, 40, 50, 60, 70, 80, 90, 100. Can you think of any more?

4.6 Dividing by sharing

> We are learning to solve division problems by sharing.

Before we start

Nuria has nine sweets and wants to share them with Isla and Finlay. Use counters and work out how many sweets they will each get.

> We can divide by sharing equally.

Let's learn

Isla picked 12 flowers and wants to share them with Amman.

She works out how many they will both get by drawing a picture …

They get 6 flowers each.

We can write a division sentence to record this:
$12 \div 2 = 6$

Amman   Isla

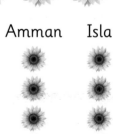

1) Use counters to help you work out these problems.

   a) Share 16 marbles between four children.
      How many marbles will each child get?

   b) Share 24 leaves between eight caterpillars.
      How many leaves will each caterpillar get?

   c) Share 30 carrots between five donkeys.
      How many carrots will each donkey get?

   d) Share 21 books between seven students.
      How many books will each student get?

   e) Share 60 fish between 10 penguins.
      How many fish will each penguin get?

2) Nuria is helping her teacher share out glue sticks.
   Work out how many each table gets if there are:
   a) 18 glue sticks and six tables
   b) 20 glue sticks and five tables
   c) 12 glue sticks and four tables
   d) 40 glue sticks and 10 tables
   Write each answer as a division sentence.

CHALLENGE! ....

Colour in the correct number of circles in the correct
number of columns to help you solve this problem:

$48 \div 8 = ?$

Resource
1B_4.6_
Challenge

4.7 Dividing by grouping

We are learning to solve division problems by grouping into equal sets.

Before we start

There are seven rows of 10 potatoes in the garden. How many potatoes altogether?

Skip count to work out the answer.

We can divide by making equal groups and skip counting to work out how many.

Let's learn

Amman has 15 marbles. He wants to put them into bags of five each. How many bags will he need?

He sorts them into equal groups and skip counts in fives to work out the answer.

| 5 | 10 | 15 |

He needs three bags.

We can record this problem as a division sentence:

15 ÷ 5 = 3

1) The children have been baking. Can you work out how many boxes they will need by skip counting? You could use counters or draw a picture to help you.

   a) Isla has made 18 muffins and wants to put two in each box.

   b) Amman has made 30 biscuits and wants to put five in each box.

   c) Nuria has made 27 brownies and wants to put three in each box.

   d) Finlay has made 40 scones and wants to put 10 in each box.

2) Finlay is having a party. He puts five sweets into each party bag. How many bags will he need for:

   a) 20 sweets?
   b) 35 sweets?
   c) 50 sweets?
   d) 65 sweets?

   You could use a number line to help you.

   Write each answer in your jotter as a division sentence.

⭐ **CHALLENGE!** ....

Isla has 100 marbles.

How many different ways can you find to sort these into equal groups?

4.8 Multiplying by 10

We are learning to multiply numbers by 10.

Before we start

Amman has £89 in £10 notes and £1 coins. How many £10 notes does he have? How many £1 coins?

We can use different materials to explore what happens when we multiply numbers by 10.

Let's learn

We could use cubes to see what happens when we multiply 4 by 10.

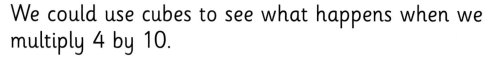

4 × 10 = 40

When we multiply by 10, the number gets 10 times bigger.

4 × 10

TENS	ONES
	4
4	0

1) Use cubes or place value materials to work out these problems:

   a) 2 × 10   b) 5 × 10   c) 7 × 10   d) 8 × 10

2) Write a multiplication sentence for each of these arrays:

   a)

   c)

   b)

3) Isla has saved money and has got £100 in £10 notes.

   How many £10 notes does she have?

   Use counters or place value materials to help you.

⭐ CHALLENGE! ....

Finlay thought of a number. He multiplied it by 10 and then doubled it and got the answer 60. What number did he start with?

Try making up more problems like this for a partner.

4.9 Dividing by 10

We are learning to divide numbers by 10.

Before we start

Nuria has planted some carrots in the garden.
Find out how many carrots there are altogether by skip counting.

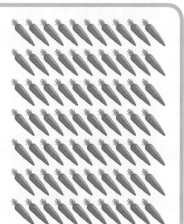

We can use different materials to explore what happens when we divide numbers by 10.

Let's learn

We could use lolly sticks to see what happens when we divide by 10. Take 30 lolly sticks and sort them into bundles of 10:

$30 \div 10 = 3$

When we divide numbers by 10 the number gets 10 times smaller.

$30 \div 10$

TENS	ONES
3	0
	3

1) Use sticks or place value materials to work out these problems by grouping them into tens:

   a) 40 ÷ 10   b) 70 ÷ 10   c) 20 ÷ 10   d) 90 ÷ 10

2) How many tens on each abacus? Write a division sentence in your jotter to show your answer.

   The first one has been done for you.

   a)
   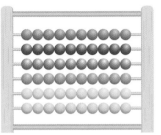

   60 ÷ 10 = 6

   b)
   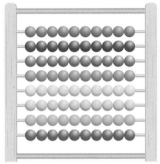

   c)

   d)
   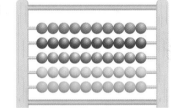

   e)
   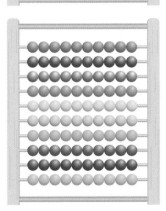

⭐ CHALLENGE! ....

a) Finlay has 10 times as many stickers as Amman. If Finlay has 140 stickers, how many stickers does Amman have?

   Use sticks or place value materials to help you.

b) Try making up more problems like this for a partner.

5.1 Making and naming equal parts

We are learning to make and name equal parts.

Before we start

Describe the girls' portions.

a) Isla's pizza

b) Nuria's pizza

A fraction is a part of a whole. When the parts are equal, we can name them.

Let's learn

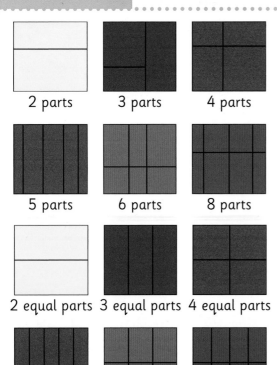

2 parts	3 parts	4 parts
5 parts	6 parts	8 parts
2 equal parts	3 equal parts	4 equal parts
5 equal parts	6 equal parts	8 equal parts

Finlay has split each square into parts:

The parts in each square are different sizes. You have not made equal parts.

Amman shows Finlay how to make equal parts by folding:

The parts are now equal. We can name these fractions.

2 equal parts: halves   **5** equal parts: fifths

3 equal parts: thirds   **6** equal parts: sixths

4 equal parts: quarters   **8** equal parts: eighths

Let's practise

1) a) Which of these have been split into equal parts?

A    B   C    D

E    F 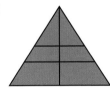   G

b) Name each of the fractions in question 1.

2) What fraction of each rectangle has been shaded blue?

a)    b) 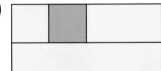   c)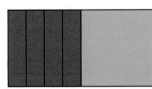

⭐ **CHALLENGE!**

You will need paper shapes (rectangles, squares, circles).

Fold the paper shapes to make a display of:

- Halves   • Sixths
- Thirds   • Eighths
- Quarters

Can we use some fractions to help us make others?

What's the easiest way to make sixths and eighths?

5.2 Identifying equal parts

We are learning to identify equal parts.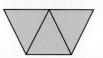

Before we start

Which of these parts can be named?

When we split a whole into equal parts, the number of parts tells us the name of the fraction.

Let's learn

Isla splits her chocolate bar into **three equal parts**.

She has made **thirds**:

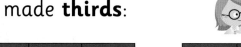

| one third | one third | one third |
| whole bar |

| one third | one third |
| two thirds |

| one third |
| one third |

She keeps **two thirds** for herself.

Amman splits his chocolate bar into **four equal parts**. He has made **quarters**:

| one quarter | one quarter | one quarter | one quarter |
| whole bar |

| one quarter | one quarter | one quarter |
| three quarters |

| one quarter |
| one quarter |

He keeps **three quarters** for himself.

The name of each fraction doesn't change when we remove parts.

1) Some of each pizza has been eaten. What fraction is left on each plate? Write your answers in your jotter.

a)

b)

c)

d)

<u>two sixths</u> of the pizza is left.

_____ of the pizza is left.

_____ of the pizza is left.

_____ of the pizza is left.

2) What fraction has been shaded? Write your answers in your jotter.

a)

b)

c)

d)

e)

f)

g)

h)

3) Draw a bar to show the following fractions (the first has been done for you):

a) Two sixths:

b) Two thirds

c) Four fifths

d) Three eighths

e) Four quarters

f) Five sevenths

Amman folds a sheet of paper to make quarters:

a)

one quarter	one quarter
one quarter	one quarter

b)

one quarter	
one quarter	one quarter

I can cut out and remove a quarter to leave three quarters.

You will need:

- Paper
- Scissors

Make the following fractions:

- Two thirds
- Three fifths
- Two sixths
- Seven eighths
- Five tenths

5.3 Naming fractions

We are learning to name fractions and understand fractional notation.

Before we start

Draw a bar model to show:
- One half
- Three quarters
- Two fifths

We can write fractions using words or numbers.

Let's learn

Nuria cuts a cake into two equal parts to make halves:

We can write this using digits:

one half = $\dfrac{1}{2}$ — Tells us **how many**

— **Names** the fraction

One of the parts is **one half**.

I have **1** out of **2** equal parts. I have **one half**.

The **2** tells us how many **equal parts** a whole has been split into.

The **1** tells us **how many of those parts** we have.

Amman has split this bar into three equal parts:

three thirds

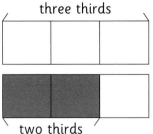

two thirds

I have made thirds.

He has coloured two out of the three thirds red:

two thirds = $\dfrac{2}{3}$ — Tells us **how many**

— **Names** the fraction

1) Write each fraction in words and numbers (the first one has been done for you):

a)

___3___ out of ___4___ parts are blue   three quarters $\frac{3}{4}$

b)

c)

d)

e)

2) Draw bars in your jotter and colour the parts to show the following fractions (the first one has been done for you):

Resource 1B_5.3_Let's_Practise_Q2

a) $\frac{4}{5}$

| one fifth | one fifth | one fifth | one fifth | one fifth |

four fifths

b) $\frac{1}{3}$

c) $\frac{3}{6}$

d) $\frac{2}{8}$

e) $\frac{4}{4}$

f) $\frac{5}{9}$

3) What fraction has not been coloured for each of your answers to question 2? (Write your answers in words and numbers)

Finlay has been given the following fraction card:

He draws bars to show it:

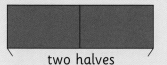

two halves   one
half

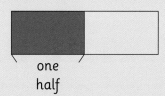
I think this says three halves.

Do you agree with Finlay?

Draw bars for the following fractions:

a) $\dfrac{5}{2}$ b) $\dfrac{4}{3}$ c) $\dfrac{7}{5}$ d) $\dfrac{6}{4}$ e) $\dfrac{8}{6}$

5 Fractions, decimal fractions and percentages

5.4 Counting fractions

We are learning to count fractions.

Before we start

Which is larger:

- One half or one sixth of a pizza?
- One third or one twelfth of a chocolate bar?
- One quarter or one eighth of all the money in Scotland?

We can count fractions using a number line.

Let's learn

Amman is making fraction cards for the class:

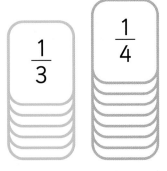

$\frac{1}{3}$ $\frac{1}{4}$

He counts the number of quarters using a number stick:

One, two, three, four, five, six, seven, EIGHT QUARTERS.

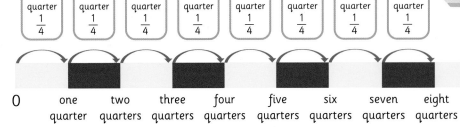

| one quarter $\frac{1}{4}$ | one quarter $\frac{1}{4}$ | one quarter $\frac{1}{4}$ | one quarter $\frac{1}{4}$ | one quarter $\frac{1}{4}$ | one quarter $\frac{1}{4}$ | one quarter $\frac{1}{4}$ | one quarter $\frac{1}{4}$ |

0   one quarter   two quarters   three quarters   four quarters   five quarters   six quarters   seven quarters   eight quarters

Nuria helps by counting out the number of thirds:

| one third $\frac{1}{3}$ | one third $\frac{1}{3}$ | one third $\frac{1}{3}$ | one third $\frac{1}{3}$ | one third $\frac{1}{3}$ | one third $\frac{1}{3}$ |

One, two, three, four, five, SIX THIRDS.

0   one third   two thirds   three thirds   four thirds   five thirds   six thirds

Let's practise

1) Copy and complete these number sticks in your jotter. Then say the counting sequence.

Resource 1B_ 5.4_Let's_ practise_Q1

   a) Quarters:

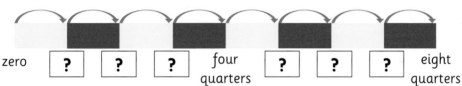

   zero   ?   ?   ?   four quarters   ?   ?   ?   eight quarters

   b) Thirds:

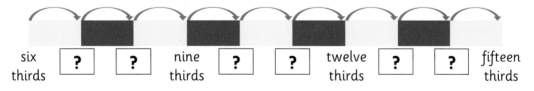

   six thirds   ?   ?   nine thirds   ?   ?   twelve thirds   ?   ?   fifteen thirds

   c) Fifths:

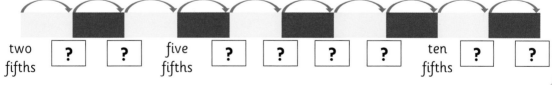

   two fifths   ?   ?   five fifths   ?   ?   ?   ?   ten fifths   ?   ?

2) Fill in the missing fractions in these number sticks.

a)

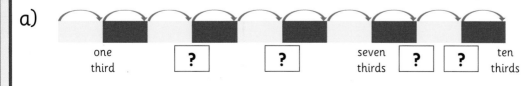

   one third   ?   ?   seven thirds   ?   ?   ten thirds

Resource 1B_ 5.4_Let's_ practise_Q2

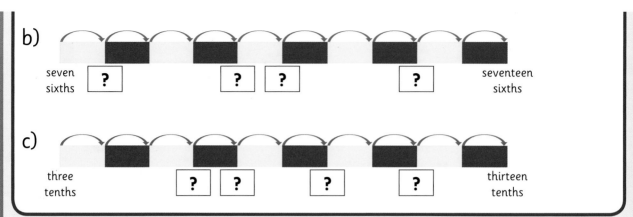

b)

| seven sixths | ? | | ? | ? | | ? | seventeen sixths |

c)

| three tenths | | ? | ? | | ? | | ? | thirteen tenths |

Number lines

3) Draw a number line to show:

   a) a count of ten jumps from zero in jumps of one seventh

   b) a count of ten jumps from six twelfths in jumps of one twelfth

   c) a count of ten jumps from ten quarters in jumps of one quarter.

CHALLENGE!

Copy the number line below in your jotter. Place the fractions in the correct positions (one has been done for you):

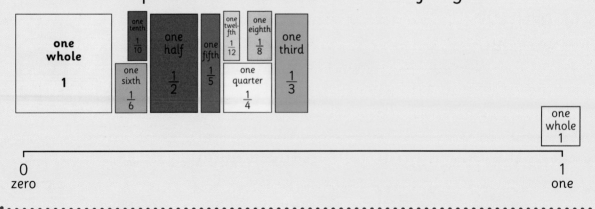

5.5 Comparing fractions

We are learning to compare fractions.

Before we start

Identify the fractions that are shaded:

a) b) c)

We can compare fractions of the same whole by looking to see which is larger or smaller.

Let's learn

You can either have one half or one third of my chocolate bar.

one whole

I can't decide. Which is bigger?

When we cut the bar in **half** there are two equal parts:

one half | one half

When we cut the bar into **thirds** there are three equal parts:

one third | one third | one third

Half the bar is **larger than** a third of the bar so Amman should choose **one half**.

What if I'd offered one half or two thirds?

Let's practise

1) **Who has more?**

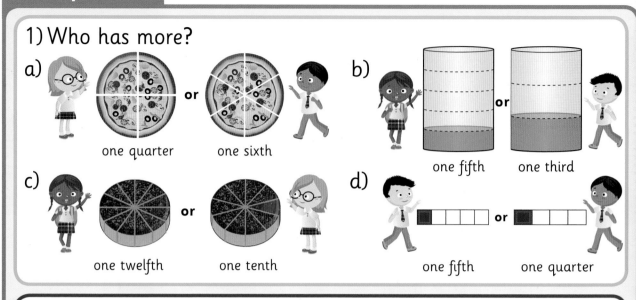

a) one quarter **or** one sixth

b) one fifth **or** one third

c) one twelfth **or** one tenth

d) one fifth **or** one quarter

2) **Which is larger:**
   a) **Three quarters** or **three twelfths** of a pizza?
   b) **Two fifths** or **two thirds** of a bottle of juice?
   c) **Four eighths** or **four tenths** of a cake?
   d) **Five sixths** or **five twelfths** of a bar of chocolate?

3) Write a sentence in your jotter to compare the following bar models. (The first one has been done for you.)

a)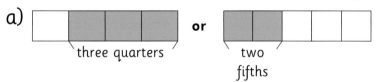

three quarters **or** two fifths

Three quarters is **larger than** two fifths of the bar.

b)

three tenths **or** four sixths

c)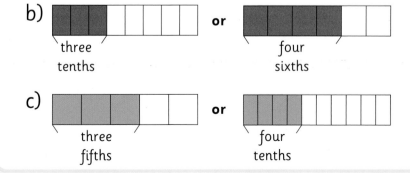

three fifths **or** four tenths

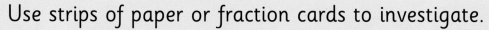

CHALLENGE!

Finlay has half a bar of chocolate.
How many fractions can you draw that are:

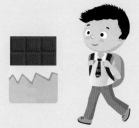

a) Smaller than one half?

b) Larger than one half?

Use strips of paper or fraction cards to investigate.

Can you find any fractions that are the same as one half?

5 Fractions, decimal fractions and percentages

5.6 Equivalent fractions

We are learning to find simple equivalent fractions.

Before we start

Which are the same?

A

B

C

D

three halves

one half and one quarter

one whole and two quarters

three quarters

We can find fractions that are equal in size.

Let's learn

Finlay has to get exactly half of a pizza:

I can cut it into two equal parts and take one of the parts.

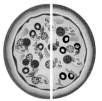 one half

You could cut it into four equal parts and take two of the parts.

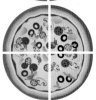 two quarters

Finlay gets exactly the same amount of food either way:

one half   two quarters

 is the same as

1) Match the portions that are equal:

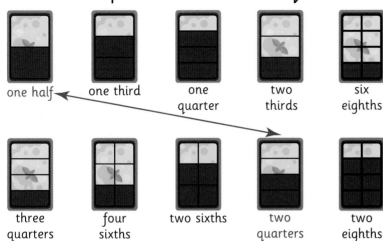

one half

one third

one quarter

two thirds

six eighths

three quarters

four sixths

two sixths

two quarters

two eighths

One half is equal to **two quarters**

2) The teacher has given each of the children the following fraction cards. Can you find matching pairs?

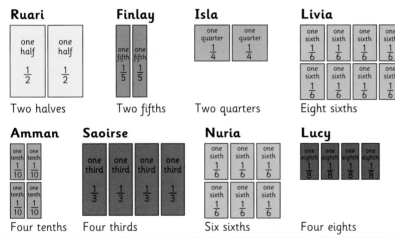

Ruari

one half $\frac{1}{2}$ one half $\frac{1}{2}$

Two halves

Finlay

one fifth $\frac{1}{5}$ one fifth $\frac{1}{5}$

Two fifths

Isla

one quarter $\frac{1}{4}$ one quarter $\frac{1}{4}$

Two quarters

Livia

one sixth $\frac{1}{6}$ one sixth $\frac{1}{6}$ one sixth $\frac{1}{6}$ one sixth $\frac{1}{6}$

one sixth $\frac{1}{6}$ one sixth $\frac{1}{6}$ one sixth $\frac{1}{6}$ one sixth $\frac{1}{6}$

Eight sixths

Amman

one tenth $\frac{1}{10}$ one tenth $\frac{1}{10}$

one tenth $\frac{1}{10}$ one tenth $\frac{1}{10}$

Four tenths

Saoirse

one third $\frac{1}{3}$ one third $\frac{1}{3}$ one third $\frac{1}{3}$ one third $\frac{1}{3}$

Four thirds

Nuria

one sixth $\frac{1}{6}$ one sixth $\frac{1}{6}$ one sixth $\frac{1}{6}$

one sixth $\frac{1}{6}$ one sixth $\frac{1}{6}$ one sixth $\frac{1}{6}$

Six sixths

Lucy

one eighth $\frac{1}{8}$ one eighth $\frac{1}{8}$ one eighth $\frac{1}{8}$ one eighth $\frac{1}{8}$

Four eights

CHALLENGE!

You will need fraction cards.

Make one whole that includes:

- Quarters and eighths
- Thirds and sixths
- Halves, quarters and eighths
- Fifths and tenths

Can you find more than one way for each?

5 Fractions, decimal fractions and percentages

5.7 Counting in fractions

We are learning to count in wholes and parts.

Before we start

Say, then fill in the missing numbers in your jotter:

a) five thirds

b) eight fifths

We can count in fractions or we can count in fractions and wholes.

Let's learn

Finlay is trying to work out the missing jumps on the number stick:

0   1   2   3

Finlay draws the number line by counting in halves:

| one half $\frac{1}{2}$ | one half $\frac{1}{2}$ | one half $\frac{1}{2}$ | one half $\frac{1}{2}$ | one half $\frac{1}{2}$ | one half $\frac{1}{2}$ |

There are two jumps between each whole number. Each jump must be one half.

0   one half   two halves   three halves   four halves   five halves   six halves

We can also fill in the number line using wholes and halves:

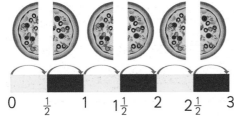

0   $\frac{1}{2}$ 1 $1\frac{1}{2}$ 2 $2\frac{1}{2}$ 3

Let's practise

1) Say, then fill in, the missing numbers:

a)

0   [?] $\frac{2}{8}$ [?]   [?]   [?]   [?] $\frac{7}{8}$ [?]   [?] $1\frac{2}{8}$

Resource 1B_ 5.7_Let's_ Practise_Q1

b)

5   [?]   [?] $5\frac{3}{7}$ [?]   [?]   [?]   6   [?]   [?]   [?]

c)

4   [?]   [?]   [?]   [?]   [?]   5   [?]   [?]   [?]   [?]

2) Copy and complete these number sticks:

a)

3   [?]   4   [?]   5   [?]   6   [?]   7   [?]   8

Resource 1B_ 5.7_Let's_ Practise_Q2

b)

5   [?]   [?]   [?]   6   [?]   [?]   [?]   7   [?]   [?]

c)

2   [?]   [?]   3   [?]   [?]   4   [?]   [?]   5   [?]

3) Draw a number sticks that count on:

a) From zero to two in jumps of one ninth.
b) From one to six and a half in jumps of one half.
c) From five to seven and a half in jumps of one quarter.

Isla has found some old broken number sticks and only some of the numbers are still visible:

Match the number lines and fill in the missing numbers.

a)
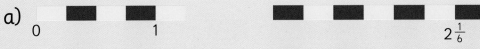

b)

c)

d)

5.8 Finding part of a set

We are learning to find part of a set.

Before we start

Find
- half of 18
- one quarter of 16

We can find a fraction of an amount by sharing equally.

Let's learn

I am getting one quarter of the cake. How many chocolate buttons should I get?

Cut the cake into quarters and share the buttons fairly.

One quarter of the cake should have two buttons.

One quarter of 8 chocolate buttons is 2 buttons.

We can show this using a bar model:

Total
8 buttons

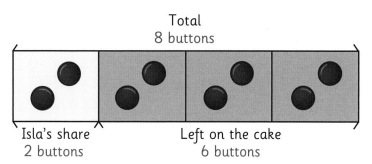

Isla's share
2 buttons

Left on the cake
6 buttons

I have taken one quarter and left three quarters of the buttons.

Three quarters of 8 chocolate buttons is 6 buttons.

1) Find the following (the first one has been done for you):

a) One half of 10 pounds:

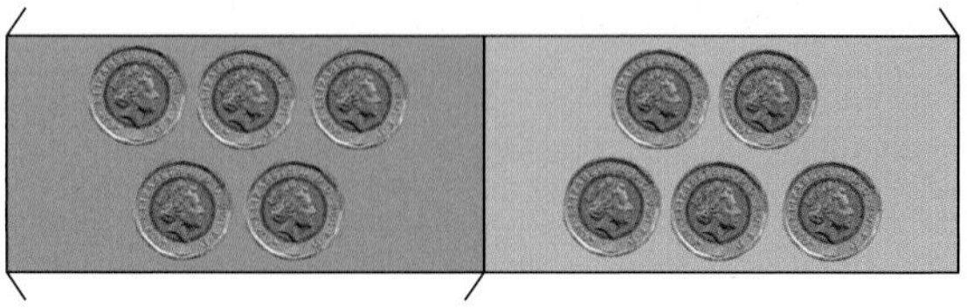

one half

One half of 10 pounds is __5__ pounds.

b) One quarter of 12 children:

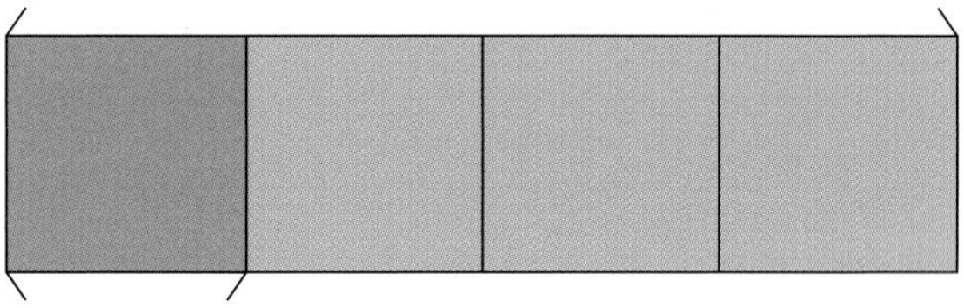

one quarter

One quarter of 12 children is __?__ children.

c) One fifth of 15 buttons:

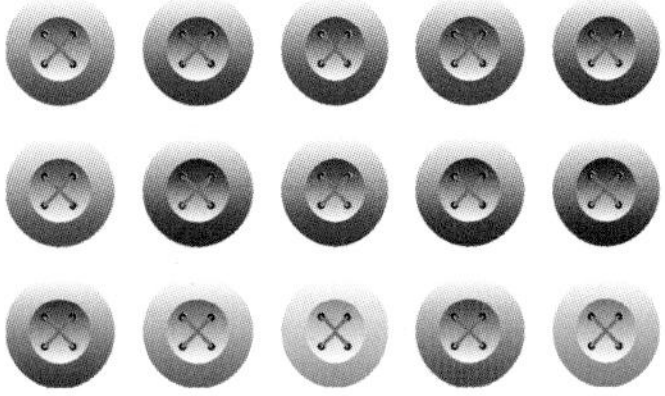

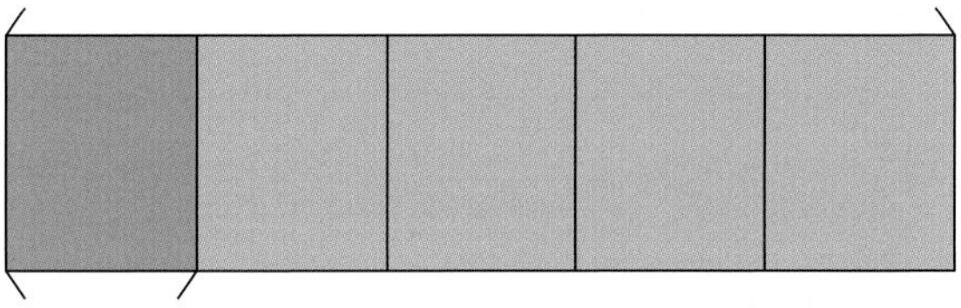

one fifth

One fifth of 15 buttons is __?__ buttons.

d) One third of 21 toy cars:

one third

One third of 21 toy cars is __?__ toy cars.

e) One sixth of 24 sweets:

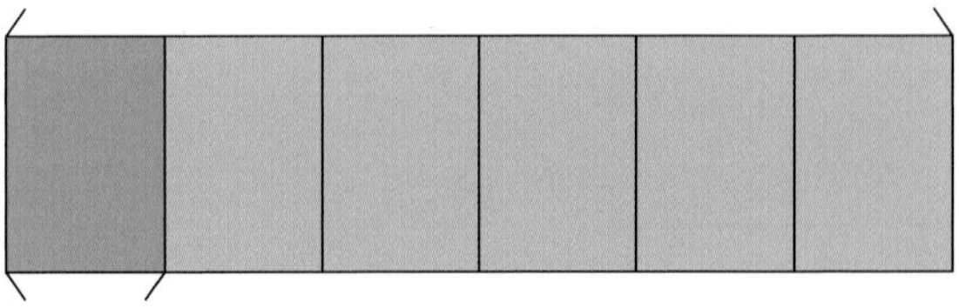

one sixth

One sixth of 24 sweets is __?__ sweets.

2) Draw bar models to solve the following:
   a) One **third** of 15 cookies = __?__ cookies
   b) One **sixth** of £18 = __?__
   c) One **quarter** of 32 children = __?__ children
   d) One **fifth** of 20 books = __?__ books

3) a) Isla had 16 sweets. She ate one quarter of them.
      How many sweets did she eat?
   b) Nuria had 24 football cards. She lost one third of them.
      How many cards did she lose?
   c) Amman had £25. He spent one fifth of his money on his
      lunch. How much was his lunch?

CHALLENGE! ....

a) Find the missing numbers:

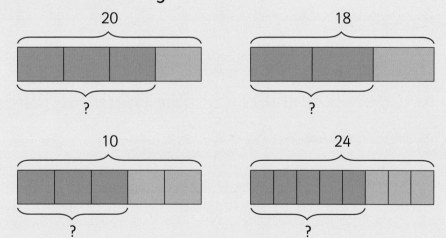

b) Make up two or three similar problems for your partner
   to solve.

6 Money

6.1 Making and recording amounts

We are learning to say how much money we have, and how to record it.

Before we start

Isla pays for this bear using this coin:
What is the smallest number of coins she can receive in change?
Draw the coins in your jotter.

13p

We can make and record amounts of money in pounds and pence.

Let's learn

This amount is £1 and 52p.
A pound is worth 100p.

So £1 and 52p is 100p + 52p and can also be written as 152p.

Let's practise

1) Write these amounts in pence in your jotter. One is done for you.
   a) £2 and 75p *275p*
   b) £4 and 21p
   c) £7 and 55p
   d) £9 and 62p

2) a) Write these amounts in £ and p. One is done for you.
      i) 620p *£6 and 20p*   ii) 310p   iii) 417p   iv) 168p
   b) Draw the smallest number of notes and coins that make each amount.

3) Write down these amounts. Record them in £ and p.

a)

b)

c)

d)

4) Write down these amounts. Record them in pence.

Write _____ p in your jotter.

a)

b)

c)

d)

e)

5) Use coins to make five different amounts. Work with a partner: one of you record them in pence and the other in £ and p. Compare your answers.

⭐ **CHALLENGE!** ....

Amman has five coins and one £5 note in his pocket.

a) What is the largest amount of money that Amman could have?

b) What is the smallest?

Explain your thinking to a partner.

c) What if all Amman's coins are different? What are the largest and smallest amounts that Amman could have now?

Explain your thinking to a partner.

6 Money

6.2 Adding amounts

We are learning to add amounts up to £1.

Before we start

a) Write this amount in two different ways.
b) Draw the same amount using the smallest possible number of coins and notes.

We add amounts to find out how much to pay, and to work out whether we have enough money.

Let's learn

Finlay has 50p. He wants to buy these two tractors.

He adds 18p + 28p = 46p.

50p is more than 46p, so he has enough money to buy both tractors.

 18p

 28p

He can work out how much money he will have left by subtracting.

50p – 46p = 4p left.

Let's practise

1) Finlay has these coins: Amman has these coins:

Nuria has these coins:  Isla has these coins:

Together, how much money do:
a) Finlay and Nuria have?   b) Amman and Isla have?
c) Finlay and Amman have?   d) Isla and Nuria have?

2) The children have 75p each. They go to a café.

Finlay wants   Nuria wants

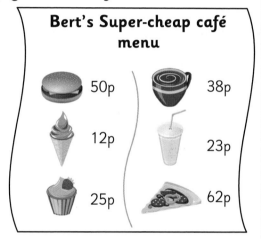

Bert's Super-cheap café menu

50p   38p
12p   23p
25p   62p

Amman wants   Isla wants

a) Work out if each child has enough to buy the items.
b) Calculate how much money they have left, if any.

3) Isla has 75p. In a charity shop, she sees a t-shirt costing 48p and a scarf costing 35p. How much more money does Isla need to be able to buy both items?

CHALLENGE! ....

Amman's mum gives him £1 to go to the shop and buy eggs, bread and milk. He can buy a sweetie with the money left over.
Bread costs 28p, eggs cost 33p and milk is 29p. A sweetie costs 8p.
Does Amman have enough money for his sweetie?

6.3 Calculating change

We are learning to work out change using the shopkeeper's method.

Before we start

What strategy would you use? Talk to a partner.

55 – 49 = ?

Before computerised tills, shopkeepers had to work out change themselves.

Let's learn

Nuria buys a sweet that costs 62p. She pays with a £1 coin.

The shopkeeper uses a counting on strategy to work out her change:

He starts at 62p and counts on. He starts with the smallest coins.

He counts out coins to the nearest 5p, then to the nearest 10p.

Then he adds on in 10p and 20p coins.

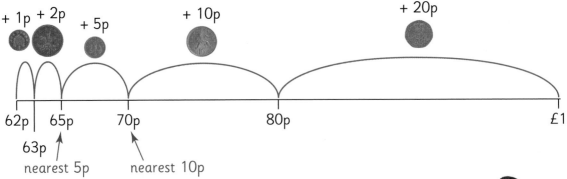

+ 1p + 2p + 5p   + 10p   + 20p

62p | 65p   70p   80p   £1
63p ↑
nearest 5p   nearest 10p

While the shopkeeper counts out loud, he puts the coins one at a time into Nuria's hand.

62p, 63, 65, 70, 80, £1

1) a) Use the shopkeeper's method to give a partner change from £1 for items costing:

    i) 70p   ii) 25p   iii) 58p   iv) 34p

  b) Draw the coins you gave your partner for each item.

2) The children have been given a £2 coin each to buy things for their pets.

Nuria buys

Isla buys

Amman buys

Finlay buys

25p   38p   65p

53p   17p   12p

30p   19p   24p

  a) Calculate the cost of the items for each child.

  b) Work out their change using the shopkeeper's method. Draw the coins.

CHALLENGE!

A shopkeeper has no 1p coins or 5p coins left in her till.

Isla wants to buy a comic which costs 49p and a bar of chocolate which costs 23p. She pays using a £1 coin.

a) Is it possible for the shopkeeper to give her the right change? Explain your thinking.

b) If so, what coins could she use?

7.1 Time word problems

> We are learning to use time facts to solve word problems.

Before we start

1) a) Copy and complete the table in your jotter to show how many days there are in different numbers of weeks.

   b) How many days are there in seven weeks?

   Write your answer in your jotter.

Days	Weeks
7	1
14	2
?	3
28	?
?	?
?	6

> Different time facts can help you solve problems.

Let's learn

There are many different time facts that you will need to learn.

Units of time build up from seconds to years as follows:

60 seconds = 1 minute

60 minutes = 1 hour

24 hours = 1 day

7 days = 1 week

12 months = 1 year

1) Finlay took four weeks to make his model aeroplane. Amman took 37 days to make his model aeroplane.

   Who made their model quicker?

2) Isla and Erin are sisters. Isla is 96 months old and Erin is 72 months old. How many years older is Isla than Erin?

3) Nuria spent 15 minutes reading her new book. She read three chapters altogether, but it took her 12 minutes to read the first two chapters.

   How long did it take her to read Chapter 3?

CHALLENGE! ....

12 weeks is the same as 78 days.

7 years is the same as 84 months.

Nuria

Amman

a) Is Nuria correct? Explain how you know. Write your answer in your jotter.

b) Is Amman correct? Explain how you know. Write your answer in your jotter.

7.2 Finding durations

We are learning to find durations in days, weeks or months.

Before we start

Copy and complete:

a) 60 seconds   =   __?__ minute

b) 60 minutes   =   __?__ hour

c) __?__ hours   =   1 day

d) 7 days   =   __?__ week

Time is not just the passing of seconds, minutes and hours, it is the passing of days, weeks and months.

Let's learn

There are many different ways you can measure the passing of time.

Shorter durations can be measured using seconds, minutes and hours.

Longer durations can be measured using days, weeks and months.

Units of time build up from seconds to years as follows:

Seconds, minutes, hours, days, weeks, months and years.

1) Finlay was going camping with his friends.
   He was away from Friday until Monday.
   How many days did his trip last?

2) Mr Irwin went on holiday on Monday the
   1st of July and returned on Sunday the
   14th of July.
   a) How many days was he away?
   b) How many weeks was he away?

JULY

Sun	Mon	Tue	Wed	Thu	Fri	Sat
	1	2	3	4	5	6
7	8	9	10	11	12	13
14	15	16	17	18	19	20
21	22	23	24	25	26	27
28	29	30	31			

3) This is a simple calendar for a year.
   a) How many months are there in
      one year?
   b) How many months would there
      be in three years?

1 Year	
January	July
February	August
March	September
April	October
May	November
June	December

CHALLENGE!

Nuria goes to piano lessons on a Tuesday and a Thursday
every week.

Her mum has paid for a block of 12 lessons.

a) How many weeks will she have piano lessons for?

b) If each lesson lasts two hours, how many hours will she
   have practised for over 12 lessons?

7.3 Recording time

> We are learning to use a variety of timers to record periods of time.

Discuss with a partner the different ways you can measure time.

How many different ways can you name?

Let's learn

> Time periods can be recorded in a variety of ways.

There are many different ways you can measure periods of time.

You will be exploring some of them on these pages.

Work with a partner on all the activities. While you complete each activity, your partner should measure how long it takes. Then you can swap roles.

Let's practise

1) a) Build five towers of 10 cubes. Your partner should march on the spot and count how many steps it takes for you to build the towers. Write the number of steps in your jotter.

   b) Now swap roles. Your partner should build five towers of 10 cubes. Count how many steps it takes for them to build it. Write your answer in your jotter.

   c) Try making 10 towers of five cubes. Does it take more or less time? Write your answer in your jotter.

2) Count how many times you can clap your hands in the time it takes to do each of these activities.

It took 24 hand claps to drink a cup of water.

a) How many hand claps does it take to say the alphabet once through? Write the tally and the answer in your jotter.

b) How many hand claps does it take to walk twice around the room? Write the tally and the answer in your jotter.

c) How many hand claps does it take to throw a ball and catch it 10 times? Write the tally and the answer in your jotter.

3) a) Use flips of a sand timer to measure how long it takes to complete these activities. Copy the table below and record your results.

Resource 1B_7.3_Let's_Practise_Q3

Activity	Number of flips of the sand timer
1) Throw a bean bag into a tray five times.	
2) Tie five knots in a piece of string.	
3) Balance on one leg for as long as possible.	
4) Bounce a ball while walking from one end of the playground to the other.	

b) Which activity took the longest amount of time? How do you know? Write your answers in your jotter.

c) Which activity took the shortest amount of time? How do you know? Write your answers in your jotter.

CHALLENGE!

Choose an activity for your partner to do. For example, they could bounce a ball on the spot or skip with a skipping rope.

What can you do in one minute?

Use a clock or a stopwatch to time one whole minute. How many did they manage to do in a minute? Now swap roles.

7.4 Telling the time

> We are learning to read half past, quarter past and quarter to.

Before we start

Discuss with a partner the different ways you can tell the time.

How many different ways can you name?

> Time can be split into whole, half and quarter hours.

Let's learn

These clocks show where the minute hand points for each time you are learning about and the digital clock number:

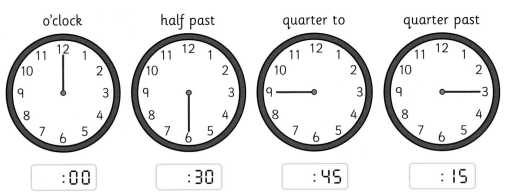

o'clock	half past	quarter to	quarter past
:00	:30	:45	:15

1) Make these times using an analogue and a digital clock face. When you are sure that you have made them correctly, draw the hands in the correct places on blank clock faces.

Blank clock faces

Example 6 o'clock

6:00

a) half past 9   b) 4 o'clock

c) half past 7   d) 1 o'clock

e) half past 2   f) 10 o'clock

2) Match up these clock faces with the correct times.

a)

b) 6: 15

c) 2: 15

| quarter past 6 |
| quarter to 10 |
| quarter past 12 |

d) (clock)   e) (clock)   f) 9:45

| quarter past 2 |
| quarter to 6 |
| quarter to 1 |

⭐ **CHALLENGE!** ....

a) Draw lines on a blank clock face where the minute hand should point:

Blank clock faces

- o'clock – blue line
- half past – red line
- quarter past – yellow line
- quarter to – green line

b) Using a clock face, make o'clock, half past, quarter past or quarter to times.

See if your partner can identify the time you have made.

7.5 Five-minute intervals

We are learning to read five-minute intervals.

Before we start

Show the hour hand and the minute hand on a clock using the digital time shown.

Blank clock faces

a)
10:30

b)
7:00

c)
9:15

d)
12:30

e)
8:45

f)
1:00

Time can be split into five minute intervals.

Let's learn

You can use your knowledge of skip counting forwards and backwards in fives when working with five-minute clock intervals.

The sequence runs forwards for the first part of the clock: o'clock, 5 past, 10 past, quarter past, 20 past, 25 past, half past, and then backwards for the second half of the clock: 25 to, 20 to, quarter to, 10 to and 5 to.

Remember to use 'quarter' for 15 minutes past and 15 to and 'half' for 30 minutes past.

1) Tell you partner the times on the clocks below:

a) b) c) d) e)

Blank clock faces

2) On a clock, show the following times:

a) Five minutes past three
b) Ten minutes to eleven
c) Twenty minutes to eight
d) Twenty-five minutes to seven
e) Quarter past five
f) Ten minutes past six

3) Write the time shown on each clock in words.

a) b) c) d) e)

CHALLENGE! ....

Isla looked at her classroom clock every quarter of an hour for two hours starting at 10 o'clock.

The pictures are muddled up.

Number the clocks from 2 to 9 in the correct order.

Resource 1B_7.5_Challenge

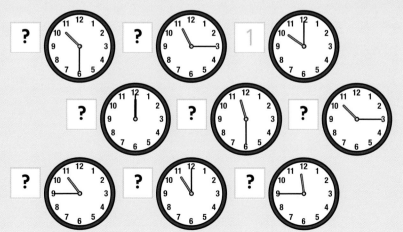

8.1 Estimating, measuring and recording length (standard units)

We are learning to estimate, measure and record length using standard units.

Before we start

Use 1 cm cubes to measure the length of these lines.

a)

b)

c)

We can measure the length or height of something using centimetres and metres.

Let's learn

This line is exactly 1 centimetre long. You can use a ruler to check.

We can write **1 cm**.

We measure short lengths using **centimetres**.

 This sharpener is 3 cm long.

1 metre is exactly
100 centimetres.

We measure longer
lengths using **metres**.

You should find a
metre stick in your
class. Check that
it is 100 cm long.

This table is
1 m tall.

1) Estimate, then measure, the length of each of the bars below.
Write your estimates and the actual lengths in your jotter.

a)
b)
c)
d)
e)
f)

2) Draw lines that are exactly:

a) 5 cm (centimetres)	in your jotter
b) 8 cm	in your jotter
c) 11 cm	in your jotter
d) 16 cm	in your jotter
e) 65 cm	in the playground with chalk
f) 80 cm	in the playground with chalk
g) 1 m (metre)	in the playground with chalk
h) 3 m	in the playground with chalk

3) In your jotter, list three items from your classroom that you could measure with a

a) ruler   b) metre stick

Estimate their length, then measure accurately.

CHALLENGE!

Find items in your classroom that you think will have a length of:

- 20 cm
- 40 cm
- 1 m
- 150 cm
- 2 m

Use a ruler or metre stick to measure the lengths accurately and check how close you were.

Try putting smaller items together to make the lengths above.

8 Measurement

8.2 Estimating, measuring and recording mass (standard units)

We are learning to estimate, measure and record mass using standard units.

Before we start

Estimate how many marbles the pencil case weighs.

We can measure the mass of an object using grams and kilograms.

Let's learn

Amman uses grams to measure the mass of his pen:

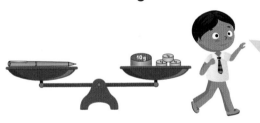

The pen balances with the weights. The pen must weigh 13 grams.

1000 lots of is equal to **1 kg**

We can use grams to measure lighter objects.

1000 grams   =   1 kilogram

We can use kilograms to measure heavier objects.

Isla wants to measure the mass of her lunchbox:

My lunchbox weights 1 kilogram and 123 grams.

1) These items balance with the weights shown.
Write down the mass of each.

a) sharpener

b) mobile

c)   school bag

d) shopping

2) For this question you will need a pan balance and weights.
Find how many you need to balance:

a) __?__ pencils balance with 100 grams.

b) __?__ cubes balance with 150 grams.

c) __?__ marbles balance with 150 grams.

d) __?__ jotters balance with 1 kilogram.

CHALLENGE! ....

You will need:

pan balance   weights   items to weigh

Resource
1B_8.2_
Challenge

a) Make a table like the one shown below to record your answers.

b) Estimate the mass of each item.

c) Measure accurately using the pan balance and weights.

Item	Estimate	Actual

8.3 Estimating, measuring and recording area (squares)

> We are learning to estimate, measure and record area using squares.

Before we start

a) Estimate how many:
- Post-it notes will fit on the cover of this book.
- Textbooks will fit on your desk.

b) Share your answers with a partner and check that you agree.

> The area of a 2D shape is the space it takes up. We can use square tiles to measure area.

Let's learn

Amman and Isla have been asked to estimate and measure the area of this rectangle:

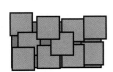

> I think 10 squares will fit. Let's put them on to see.

> That doesn't work. There can't be any gaps or overlaps. It must be done neatly like this.

The children can fit eight squares on top of the rectangle.

The **area** of the rectangle is **8 square tiles**.

You can use squared paper or cubes like these if you don't have square tiles.

Squared paper

1) Estimate how many 1 cm squares will fit these shapes:

a)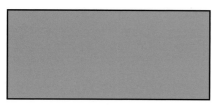

b)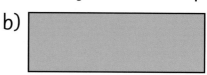

c)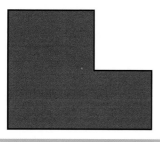

d)

e)

f)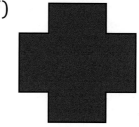

2) Use squared paper or cubes to measure the area for each of the shapes in question 1.

3) Draw shapes you estimate to have an area of:

a) 6 squares   b) 10 squares   c) 15 squares
d) 24 squares   e) 30 squares

Check using cubes or squared paper.

CHALLENGE! ....

Amman has made a shape with an area of 12 squares:

Use squared paper to find as many different shapes as you can with the same area.

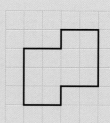

Squared paper

8 Measurement

8.4 Estimating, measuring and recording capacity

We are learning to estimate, measure and record capacity using standard units.

Before we start

a) Estimate how many glasses could fill a bottle of juice.

b) Estimate how many cups could fill a kettle.

We can measure the capacity of a container using millilitres and litres.

Let's learn

Capacity is the total amount a container can hold.

This glass can hold 200 millilitres of water.

Volume is the amount held by the container.

The glass contains 150 millilitres.

We can measure smaller volumes of liquid using millilitres.

The tablespoon holds 10 millilitres of water.

The can holds 330 millilitres of juice.

We can write **10 ml** and **330 ml**.

We can measure larger volumes of liquid using litres:

  This measuring jug holds one litre of water.

  The kettle holds three litres of water.

We can write **1 l** and **3 l**.

1) Estimate which will hold more or less than a litre:

a) bucket

b) jam jar

c) yoghurt pot

d) kettle

e) watering can

f) water bottle

g) fish tank

h) cup

2) What unit of measurement would you use to measure: litre or millilitre?

a) egg cup

b) petrol can

c) juice carton

d) water pistol

e) teaspoon

f) bath

g) water cooler

h) cooking pot

3) You will need:

Supermarket fliers or access to the internet.

Find the capacity of five different containers and record these in litres or millilitres.

CHALLENGE!

You will need:

- measuring jug
- 2-litre bottle
- four or five different-sized drinks bottles (with labels on)
- funnel

capacity	estimate	actual

Find and record the capacity of each drinks bottle.

How many drinks bottles will it take to fill the 2-litre bottle? (Try estimating first.)

8.5 Converting standard units of length/height

We are learning to convert standard units of length/height (cm, m).

Before we start

How long is each line?

a)

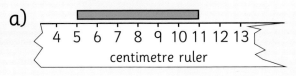

```
4  5  6  7  8  9  10 11 12 13
        centimetre ruler
```

b)

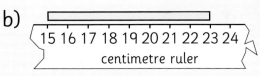

```
15 16 17 18 19 20 21 22 23 24
        centimetre ruler
```

We can state length/height using centimetres, metres or both.

Let's learn

'Cent' means 'one hundred', so one metre is the same as one hundred centimetres.

This is a metre stick. It is exactly 1 metre long.

Look at a metre stick closely. You will see that it is 100 centimetres long.

```
|——————————————— 1 metre ———————————————|
  10    20    30    40    50    60    70    80    90   100
```

The van is 2 metres tall.

1 metre = 100 centimetres or **1 m = 100 cm**

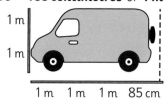

1 m
1 m

1 m 1 m 1 m 85 cm

2 m = 200 cm
3 m 85 cm = 385 cm

We can also say that it is 200 centimetres tall.

The van is 3 metres and 85 centimetres long.

We can also say that it is 385 centimetres long.

Let's practise

1) Write down how tall each object is in metres and centimetres:

a)

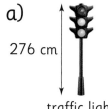

276 cm

traffic light

b)

175 cm

car

c)

107 cm

moped

d)

238 cm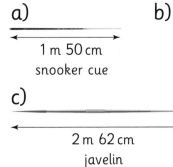

ambulance

e)

304 cm

bus

f)

460 cm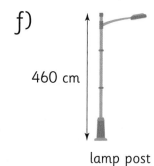

lamp post

2) Convert each of the measurements into centimetres:

a)

1 m 50 cm
snooker cue

b)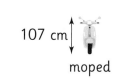

2 m 9 cm
fishing rod

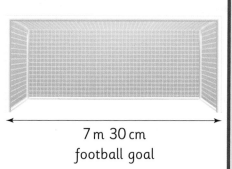

c)

2 m 62 cm
javelin

d)

1 m 35 cm
golf club

e)

7 m 30 cm
football goal

You will need:

1 metre

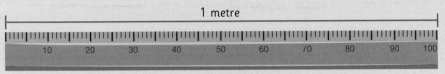

metre stick

Find five items that you estimate to be over 1 metre.
For example:

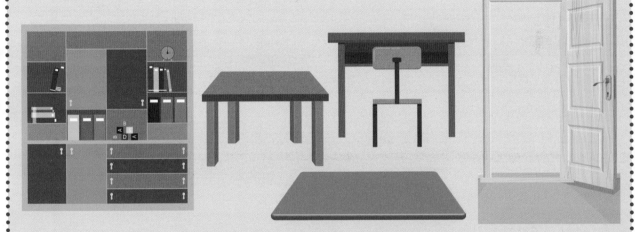

With a partner, measure the length of each item.
Write down the measurements in:

- metres and centimetres
- centimetres

Item	metres and centimetres	centimetres

8.6 Converting standard units of mass

We are learning to convert standard units of mass.

Before we start

What is the mass of each item?

a)

b)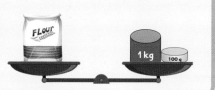

We can state mass using grams, kilograms or both.

Let's learn

 This is a kilogram weight. The potatoes weigh one kilogram.

 This is a gram weight. The potatoes weigh one thousand grams.

1000 1 gram weights

'Kilo' means 'one thousand', so one kilogram is the same as one thousand grams.

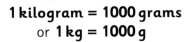

The bucket weighs 1 kilogram 234 grams.

1 kilogram = 1000 grams
or 1 kg = 1000 g

1 kg 234 g

Let's practise

1) Write each measurement. The first one has been done for you.

a)

  <u>1</u> kg <u>535</u> g

b)
  <u>?</u> kg <u>?</u> g

c)

  <u>?</u> kg <u>?</u> g

d)

  <u>?</u> kg <u>?</u> g

e)

  <u>?</u> kg <u>?</u> g

f)

  <u>?</u> kg <u>?</u> g

2) Draw the weights that would be needed to balance the scales. The first one has been done for you.

a)

  254 g

b)

  1kg 307 g

c)

  2kg 32 g

d)
  1kg 9 g

e)

1kg 506g

f)

3kg 70g

Resource
1B_8.6_
Challenge

CHALLENGE! ....

You will need a pan balance and weights.

Find five items that you estimate to be over 1 kilogram.

For example:

With a partner, measure the mass of each item.

Write down the measurements in:

- kilograms and grams
- grams

Item	kilograms and grams	grams

8 Measurement

8.7 Reading scales – length

We are learning to read scales on measuring devices (length).

Before we start

Draw lines that are: 5 cm; 8 cm; and 20 cm long.

We can use a ruler or measuring tape to measure length accurately.

Let's learn

Amman is trying to measure the length of this line:

I think the line is 10 cm long because it stops at the 10.

That's not right. You need to make sure that it is in line with the zero to begin with.

Let's try again...

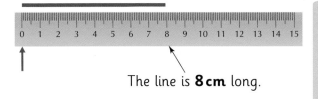

The line is **8 cm** long.

Now it is lined up with the zero we can measure the length accurately.

Finlay is trying to measure the length of the piece of string:

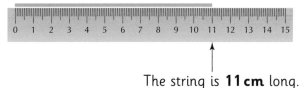

I think the string is 9 cm long.

That's not right. You need to make sure that the string is straight.

Let's Try again..

Now it is straight we can measure the length accurately.

The string is **11 cm** long.

Let's practise

1) The children are trying to measure different lengths:

The chocolate bar is 12 cm long.

The purse is 9 cm long.

The pen is 10 cm long.

The worm is 7 cm long.

a) Who is making an accurate measurement?

b) What advice would you give to the others?

2) The children use a tape measure to measure larger lengths. Write the length of each in your jotter (the first one has been done for you):

a)

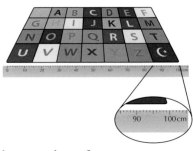

b)

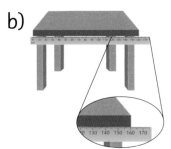

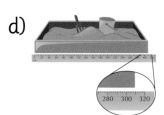

Length of carpet = <u>100 cm</u> or <u>1 m</u>

c)

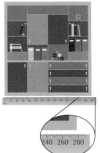

d)

CHALLENGE!

You will need:

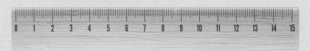

ruler   measuring tape

Resource
1B_8.7_
Challenge

Find items that you estimate to be between:

- 10 cm and 20 cm
- 20 cm and 30 cm
- 30 cm and 50 cm

- 50 cm and 1 m
- 1 m and 2 m
- 2 m and 5 m

Copy and complete the table below with your measurements:

Between	item	length
10 cm and 20 cm		
...		

8.8 Reading scales – mass

We are learning to read scales on measuring devices (mass).

Before we start

Write the mass of these items:

We can use scales to measure mass accurately.

Let's learn

We can use a pan balance and weights to measure the mass of an object:

We can also use different types of scales that don't need any weights. We can read the scale to tell us the mass.

Finlay measures his weight using bathroom scales:

Nuria measures the weight of the bag of sugar using kitchen scales:

I weigh 25 kilograms.

The sugar weighs 1 kilogram.

Let's practise

1) What is the mass of the following? Write your answers in your jotter.

a)

b)

c)

d)

e)

f)

2) Decide what type of scale you would use to weigh the following and draw the display:

a)

18 kg

b)

3 kg

c)

70 kg

d)

7 kg

e)

9 kg

f) CEMENT

36 kg

3) Weigh the following using suitable scales and record their mass (to the nearest kilogram):

a) a school bag

b) yourself then a friend

c) a box of books

d) a tray of pens

e) a bag or tray of filled water bottles

⭐ **CHALLENGE!** ....

You will need: suitable scales, a sturdy bag or box, various items to weigh.

Fill the bag (or box) with items you estimate to weigh the following:

- 1 kg
- 2 kg
- 4 kg
- 8 kg
- 10 kg

a) Weigh the box to see how close you were.

b) Try to get the exact weight by adding or removing items.

8.9 Reading scales – capacity

We are learning to read scales on measuring devices (capacity).

Before we start

What are the measurements for both of the following:

a)

b)

We can use a measuring jug to measure volumes of liquid accurately.

Let's learn

Finlay and Nuria have been asked to find out how much water is left in the water bottle:

The capacity of the measuring jug is 1 litre. The scale on the side tells us the volume of the water inside.

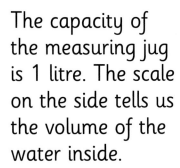

We can measure volume accurately using a measuring jug.

The water level is in line with 600. The water bottle contained 600 ml of water.

The children are making pancakes and need to measure out 500 ml of milk to add to the mixture. They can use the measuring jug to measure out the exact amount.

1) What volume of water is contained in each of the measuring jugs? (the first one has been done for you):

a)

Volume: **500 ml**

b)
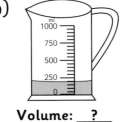

Volume: __?__

c)

Volume: __?__

d)

Volume: __?__

e)

Volume: __?__

f)

Volume: __?__

g)

Volume: __?__

2) Use a measuring jug to get the following volumes of water:

a) 300 ml   b) 450 ml   c) 900 ml   d) 20 ml

3) Make a sketch of each of the measurements in question 2.

Work with a partner. You will need:

- a 1 l measuring jug
- some different containers (units of measure)
- sticky labels (or post-its)

My estimate ← → My partner's estimate

For each container:

- Fill with water.
- Estimate the level on the measuring jug (mark with a sticky label).
- Pour the water into the measuring jug to check.

How much water does each container hold?

Whose estimate was the closest to the actual measurement?

9 Mathematics, its impact on the world, past, present and future

9.1 Ways numbers are used in work

We are learning ways numbers are used in work.

Before we start

With a partner, discuss where and how numbers are used to help people do their job.

Numbers are used in everything we do.

Let's learn

People use numbers at work every day, to help them do their jobs.

Some examples of the ways maths is used include:

· Increasing a recipe.

· Working out how much to charge for something.

· Measuring medicine doses.

· Making sense of statistics and graphs in the news.

1) Look at the following people and discuss how they use numbers in their jobs. Make a list with a partner.

2) List the different ways numbers would be used when baking a loaf of bread. Use the picture to help.

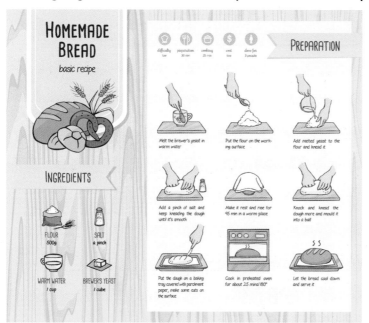

HOMEMADE BREAD
basic recipe

INGREDIENTS

FLOUR
500g

SALT
a pinch

WARM WATER
1 cup

BREWER'S YEAST
1 cube

PREPARATION

Melt the brewer's yeast in warm water

Put the flour on the working surface

Add melted yeast to the flour and knead it

Add a pinch of salt and keep kneading the dough until it's smooth

Make it rest and rise for 45 min in a warm place

Knock and knead the dough more and mould it into a ball

Put the dough on a baking tray covered with parchment paper, make some cuts on the surface

Cook in preheated oven for about 25 mins at 180°

Let the bread cool down and serve it

⭐ CHALLENGE!

Numbers help us build things.

Talk to a partner about this statement.

Imagine you are building a house. Make a list of the numbers and mathematics you would use.

10.1 Investigate repeating patterns

We are learning to investigate repeating patterns.

Before we start

Copy and draw the next two shapes in these patterns in your jotter:

a)

b)

Shapes, symbols and movements can be **repeated** to form a pattern or sequence.

Let's learn

Patterns and **sequences** can be made up with shapes, symbols and movements that **repeat**.

Repeating shape pattern

Repeating symbol pattern

Repeating rotation pattern

1) Copy and complete these patterns in your jotter:

a) ? ? b)  ? ?

2) Copy and complete these patterns in your jotter, filling in the gaps by adding the missing elements:

a)   ?   ?

b) ?   ?   ?

c) ?   ?

3) Copy and complete these patterns, filling in the gaps by adding the missing elements:

a) ?   ?   ?

b)   ?   ?

c) ...

CHALLENGE!

Use these shapes to create some repeating patterns of your own. You may use each shape as often as you wish.

10 Patterns and relationships

10.2 Recognise and continue number sequences

> We are learning to recognise and continue number sequences.

Before we start

Look at these numbers.
Tell a partner what pattern you see.

> Numbers can create patterns.

Let's learn

Patterns are not only created by shapes, symbols and movement.

Numbers also create patterns.

Use skip counting to help you to recognise how the number patterns are increasing.

1) Copy and complete these patterns. Draw the next group of circles in your jotter. Write the missing numbers.

a)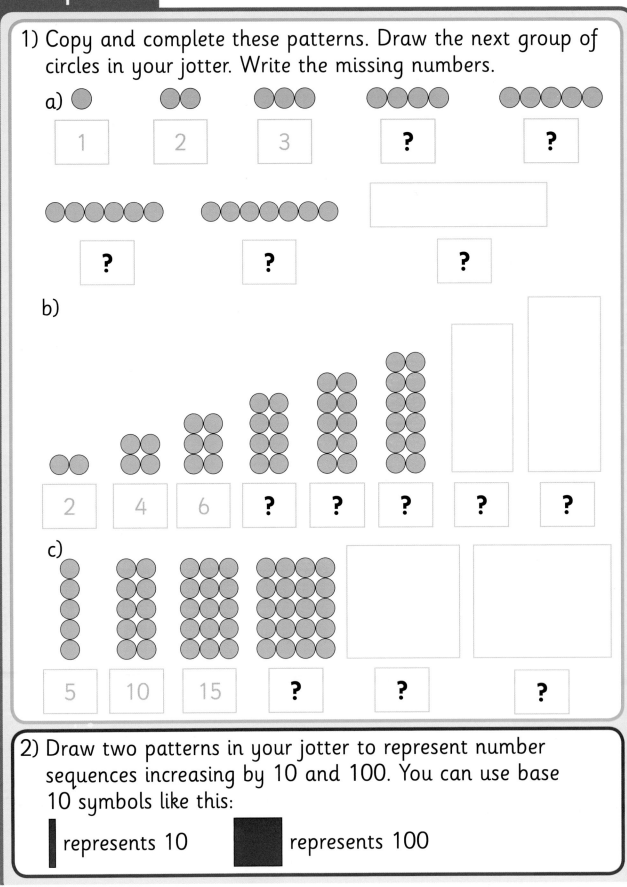

1   2   3   ?   ?

?   ?   ?

b)

2   4   6   ?   ?   ?   ?   ?

c)

5   10   15   ?   ?   ?

2) Draw two patterns in your jotter to represent number sequences increasing by 10 and 100. You can use base 10 symbols like this:

| represents 10   ■ represents 100

3) Continue these number patterns:

a) | 1 | 2 | 3 | 4 | 5 | 6 | 7 | ? | ? | ? |

b) (2) (4) (6) (8) (10) (12) (14) (?) (?) (?)

c) 5 10 15 20 25 30 ? ? ?

d) 10 20 30 40 50 60 ? ? ?

e) | 100 | 200 | 300 | ? | ? | ? |

CHALLENGE!

Can you complete these sequences?
Copy the number sequences and write your answers in your jotter.

a) 12, 14, 16, 18, ?, ?, ?

b) 15, 20, 25, 30, ?, ?, ?

c) 60, 70, 80, ?, ?, ?, ?

d) 22, 24, ?, 28, 30, ?, 34.

e) 45, ?, 55, 60, ?, 70, 75.

11.1 Solve simple equations

> We are learning to solve simple equations.

Before we start

Find the value of the following shapes:

1) ⭐ + 4 = 20

2) 2 × 8 = ⬟

3) 13 + 7 = ◼

4) 20 − 9 = ●

> **Equations** are number sentences that must balance.

Let's learn

Equations are like balancing scales.

The amount on one side must **equal** the amount on the other side.

1) If = 5 and = 10, what do you need to add to balance the scales?

a)

b)

c)

2) Copy and complete the following number sentences in your jotter:

a) 16 + ? = 20

b) 30 − ? = 15

c) ? + 19 = 25

d) 13 + ? = 30

CHALLENGE!

 = 3, = 2 and = 5

a) Place the shapes onto the scales, making sure each side equals the same amount.

b) Can you find one more way to solve the problem?

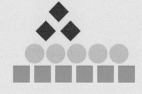 

12 2D shapes and 3D objects

12.1 Properties of 2D shapes

> We are learning to sort and describe 2D shapes.

Before we start

Amman says only one of these is a right angle.

Do you agree?

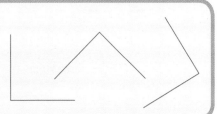

> We can describe 2D shapes by the number of sides and the number of corners.

Let's learn

Some shapes have square corners.
Square corners are **right angles**.

We can check if a corner is a right angle using the corner of a piece of paper.

Let's practise

1) Sort the shapes. Write the names of the shapes in the correct box. Some shapes will be in more than one box.

   (Use resource 1B_12.1_Let's_Practise_Q1 or copy the table in your jotter.)

Resource 1B_12.1_Let's_Practise_Q1

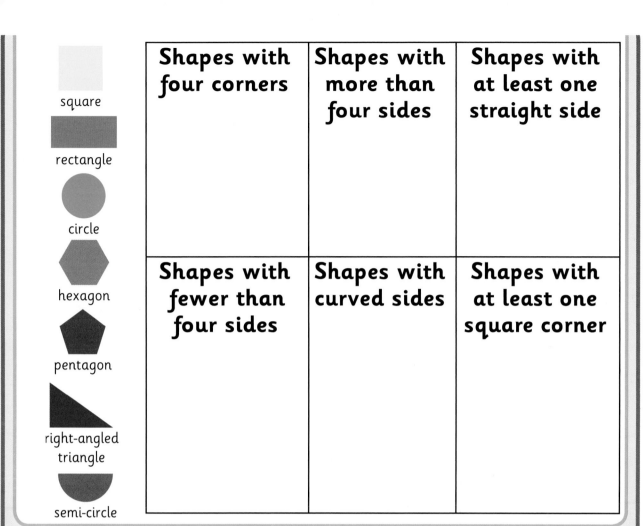

square	Shapes with four corners	Shapes with more than four sides	Shapes with at least one straight side
rectangle			
circle			
hexagon	Shapes with fewer than four sides	Shapes with curved sides	Shapes with at least one square corner
pentagon			
right-angled triangle			
semi-circle			

2) Use geoboards to make **different** shapes.

At least one of your shapes must have a corner that is a right angle.

a) Make two shapes with:

  i) three sides   ii) four sides

  iii) five sides   iv) six sides

These triangles both have three sides but only the lower one has a right angle.

b) How many corners do your shapes have?
   Copy and complete the table:

Shapes	Number of sides	Number of corners
Triangles	3	
Rectangles and squares	4	

Pentagons	5	
Hexagons	6	

c) What can you say about the number of sides and corners?

3)

You have groups of two hexagons, four triangles or three rectangles.

a) Which group has more sides?

b) Which group has more corners?

c) Which group has more right-angled corners?

⭐ **CHALLENGE!** ....

Nuria is making shapes out of matchsticks. She has used seven matchsticks to make these two squares.

She has 20 matchsticks altogether.

Use matchsticks or lolly sticks to help you.

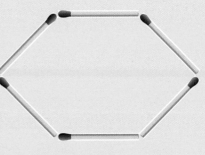

a) What is the largest number of these squares she can make?

b) What if Nuria makes hexagons? What is the largest number of hexagons she can make?

Hint: try building the shapes so they share as many matchsticks as possible.

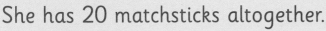

12.2 2D shapes in 3D objects

> We are learning to describe and sort 3D objects.

Before we start

The cat has sat on some shapes.
What might the shapes be?
Talk to a partner.

> Their faces help us to describe 3D objects. The faces are 2D shapes.

Let's learn

The faces of a cube are all **square**.
Cubes have 12 edges and 8 corners.

Let's practise

1) Write down **two** things that are **the same** and **one** thing that is **different** about these pairs of objects.

a)

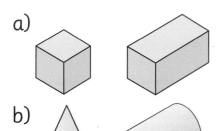

> Think about faces, edges and corners.

b)

2) How many of each 2D shape would you need to make each 3D object?

Copy and complete the table.

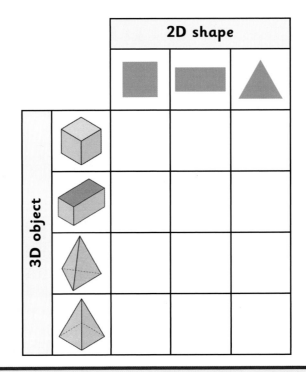

3) Nuria is describing 3D objects by talking about their faces. Name each object she describes.

a) It has two circular faces.

b) It has no flat faces.

c) It has six rectangular faces.

d) It has exactly one circular face.

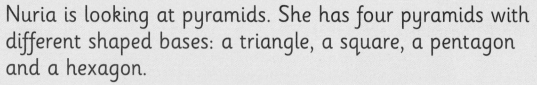
Nuria is looking at pyramids. She has four pyramids with different shaped bases: a triangle, a square, a pentagon and a hexagon.

Resource 1B_12.2_ Challenge

a) Copy and complete the table to investigate the number of triangular faces and the number of edges on pyramids with different bases.

Shape of pyramid base	Number of triangular faces	Number of edges
Triangle		
Square		
Pentagon		
Hexagon		

b) Can you say how many edges a pyramid has from the shape of its base?

c) If a pyramid's base has 20 sides, predict how many triangular faces it would have.

12.3 Recognising 3D objects

> We are learning to recognise 3D shapes from different points of view.

Before we start

Sort these objects into two groups by writing their names.

Show a partner how you have sorted them.
Can your partner say what your sorting rule was?

> 3D objects look different from different points of view.

Let's learn

You can hold a 3D object in your hands and turn it around to see the different views.

Let's practise

1) Here are some 3D objects, drawn as if you are looking at their base.

A   B ●   C ▭   D ▲

Here are the same objects drawn from a different point of view.

1   2   3   4

Match the 3D objects to their base view by copying and completing the table.

Base (letter)	A	B	C	D
Object (number)				

2) In each group of three 3D objects, two are the same and one is different.

Write down the letters of the objects that are the **same**.

a)
   A   B   C

b)
   A   B   C

c)
   A   B   C

3) Look at these 3D objects. They have all been drawn as if we are looking down on them from above.

Can you identify each object? Write your answers in your jotter.

a)   b)   c)   d)

CHALLENGE!

These diagrams show a 3D object made of four cubes.

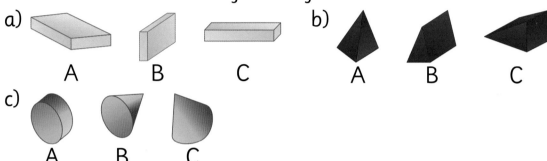

Three of the diagrams show the same 3D object, seen from different points of view.

One of the diagrams shows a different 3D object.

Can you find the odd one out? You could try making the objects out of cubes.

12.4 Tiling patterns

We are learning to identify 2D shapes that tile.

Before we start

Amman says he is thinking of a 2D shape with six corners and five sides. What is Amman's shape? Can you draw it?

Tiling is when a surface is covered with flat shapes, without any overlaps or gaps.

Let's learn

Most people have tiles in their kitchen or bathroom.

These are usually simple tiling patterns, with one shape that repeats over the whole surface.

In some tiling patterns, like this one, the sides and the corners all touch.

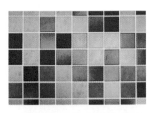

In some tiling patterns, the corners do not all touch.

Let's practise

1) Where have you seen repeating patterns of shapes?

   Write down where you can see tiling patterns.
   If you have a camera, take some photographs of tiling patterns.

2) Predict which of these 2D shapes will tile so that the sides and corners all touch.

   Test your predictions by making tiling patterns, using a 2D shape to draw around.

   Remember, the shapes must not overlap and there must be no gaps.

   a)   b)   c)   d)

3) These shapes are made up of squares.

   Will they all tile? Make tiling patterns to find out.

CHALLENGE!

A designer has been asked to tile a floor using tiles like this one.

The tile is a rectangle, where the long side is exactly twice the length of the short side.

The designer has made this pattern:

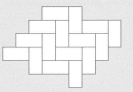

 Find out the name of this tiling pattern.

Can you help the designer by making **at least three** different tiling patterns she can use?

13 Angles, symmetry and transformation

13.1 Making turns

> We are learning to make full, half and quarter turns.

Before we start

Write down at least two different ways the Big Bad Wolf can get to the Little Pig's house of straw.

Use left, right, forward and backward.

> We make turns to face in a different direction.

Let's learn

Turns can be **clockwise** or **anti-clockwise**.

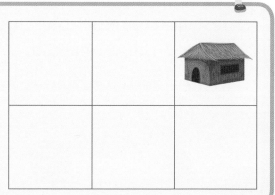

anti-clockwise   clockwise

If we make a full turn, we end up facing in the same direction we started:

start   finish

If we make a half turn, 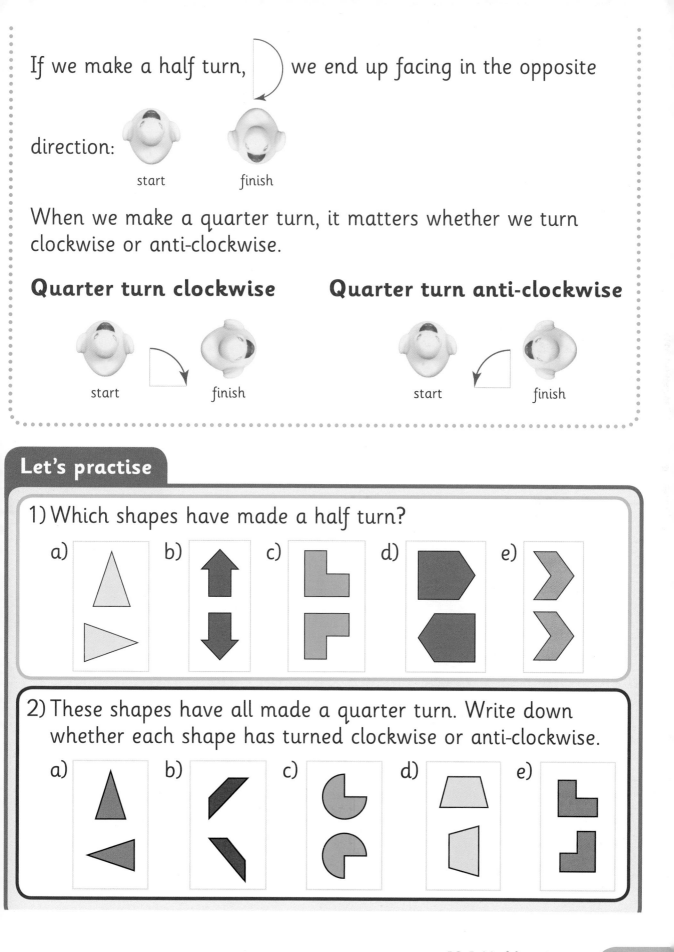 we end up facing in the opposite

direction:

start   finish

When we make a quarter turn, it matters whether we turn clockwise or anti-clockwise.

Quarter turn clockwise

start   finish

Quarter turn anti-clockwise

start   finish

Let's practise

1) Which shapes have made a half turn?

   a)   b)   c)   d)   e)

2) These shapes have all made a quarter turn. Write down whether each shape has turned clockwise or anti-clockwise.

   a)   b)   c)   d)   e)

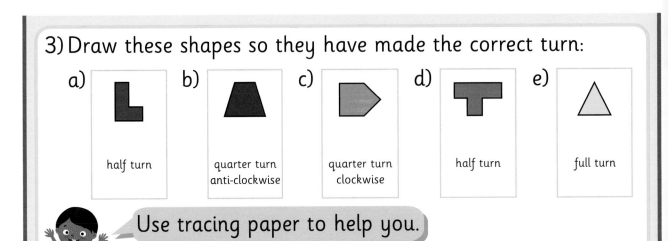

3) Draw these shapes so they have made the correct turn:

a) [L shape] half turn

b) [trapezium shape] quarter turn anti-clockwise

c) [pentagon/arrow shape] quarter turn clockwise

d) [T shape] half turn

e) [triangle shape] full turn

Use tracing paper to help you.

⭐ **CHALLENGE!** ....

Finlay is on his bike. He starts in this position.

As he cycles, he makes a quarter turn clockwise, then a half turn, another quarter turn clockwise, another half turn and then a quarter turn anti-clockwise.

Which direction does he end up facing?

a)    b)    c)    d) [bike facing left]

Hint: You could stand up and turn yourself.
Ask a partner to read the directions.

13.2 Right angles

We are learning to find right angles.

Before we start

This arrow has been turned a quarter turn clockwise. Draw the position the arrow started in.

An angle measures the amount of turn. Angles are measured in **degrees.**

Let's learn

Angles can be different sizes. Here are some different angles.

A quarter turn has a special name. It is called a **right angle** and it measures **90 degrees**.

90 degrees

1) Make a right-angle measurer from the corner of a piece of paper.

Look around your classroom and find things that have right angles. Make a list of all the things that have a right angle in your jotter.

2) Use a right-angle measurer to find the shapes that have right angles in them.

a)   b)   c)   d)

e)   f) 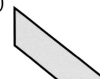   g)   h)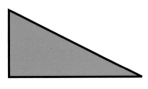

3) Find all the right angles in this picture.
Mark them like this:

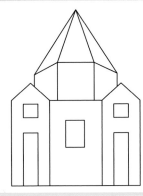

Resource
1B_13.2_Let's_Practise_Q3

CHALLENGE!

Can you work out how many degrees there are in a half turn?

How many degrees are in a full turn?

13 Angles, symmetry and transformation

13.3 Grid references

We are learning to use grid references to locate objects and places.

Before we start

Which of these birds are the same?

A B C D E F G H

Grid references are used to find places on a map or objects on a grid.

Let's learn

Grid references are always read the same way: horizontally then vertically.

horizontal then vertical

Think about walking along the corridor, then going up in the lift.

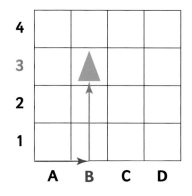

The triangle is at B3.

1)

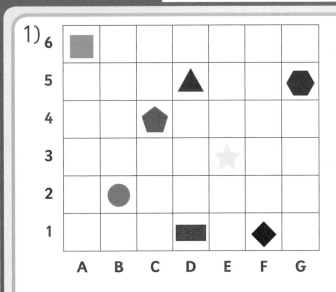

Write down the grid references for each shape in your jotter.

a) ▲   b) ●

c) ⬡   d) ⬠

e) ◆   f) ★

g) ■   h) ▬

2) The children are at a Safari Park.

a) Write down the grid references to help them find these animals.

i) rhino

ii) cheetah

iii) hippo

iv) camel

v) antelope

vi) gorilla

b) Which animal is at
   i) F1   ii) A7   iii) H3   iv) G5?

CHALLENGE! ....

Play a game of battleships with a partner.
Each player has their own grid like this one.
Full instructions are included on resource 1B_13.3_Challenge.

Resource 1B_13.3_Challenge

Each of you has five battleships:

Mark your battleships on your grids without your partner seeing.

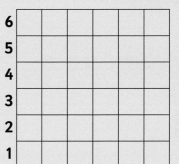

You can put your battleships anywhere on the grid, but **they must not touch**, even at the corners.

You can position them horizontally or vertically.

13.4 Symmetry

We are learning to recognise symmetry.

Before we start

Which square or squares can be coloured red to make this picture symmetrical?
Talk to a partner.

Many things around us are symmetrical.

Let's learn

Hold a mirror upright on the dotted line.

When you look in the mirror the owl looks the same as without the mirror.

This means it is symmetrical.

Let's practise

1) Use your mirror to test each object to see if it is symmetrical.
   The mirror line must go down the middle of the object.
   For each object, write yes if it is symmetrical, no if it is not.

   a)  ?

   b) ?

   c)  ?

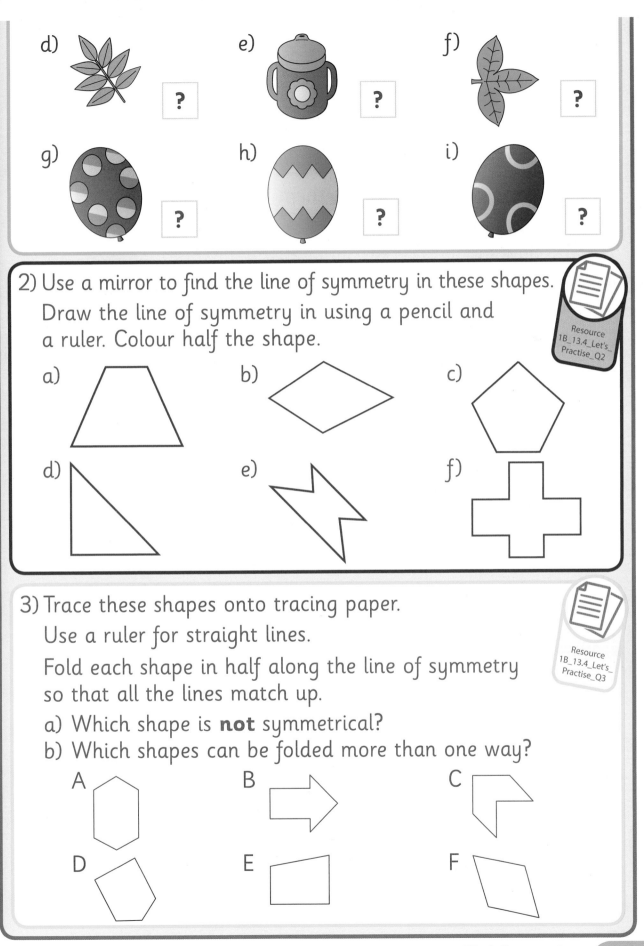

d) ?

e) ?

f) ?

g) ?

h) ?

i) ?

2) Use a mirror to find the line of symmetry in these shapes.
Draw the line of symmetry in using a pencil and
a ruler. Colour half the shape.

Resource
1B_13.4_Let's_
Practise_Q2

a)

b)

c)

d)

e)

f)

3) Trace these shapes onto tracing paper.
Use a ruler for straight lines.
Fold each shape in half along the line of symmetry
so that all the lines match up.

Resource
1B_13.4_Let's_
Practise_Q3

a) Which shape is **not** symmetrical?
b) Which shapes can be folded more than one way?

A

B

C

D

E

F

Work with a partner.

You will need interlocking square tiles, squared paper, a ruler and a red pencil.

- Take five tiles each and use them to make a T shape.

- Find how many different symmetrical shapes you can make by joining your two T shapes together.

- Record each new shape you make on squared paper.

- Mark the line or lines of symmetry with a dotted red line.

14 Data handling and analysis

14.1 Surveys and questionnaires

We are learning to collect data.

Before we start

1) Look at these images.
   Use tally marks to find the total of each fruit.

2) Use the tally marks to answer:
   a) Which fruit is the most popular?
   b) Which fruit is the least popular?

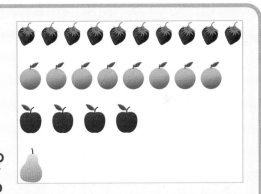

Data can be gathered in different ways.

Let's learn

Data is information that can be shared by making a graph, table or chart.

Data can include facts in words, numbers or a mixture of both.

If you want to find out what people think about something you can run a survey and collect data.

Let's practise

1) Look at the survey data below and answer the following questions in your jotter:

   a) Does Finlay like ice-cream?
   b) What is his favourite flavour?
   c) Does he like a tub or a cone?

Ice-cream survey

....

Do you like ice-cream?

☒ Yes   ☐ No

What is your favourite flavour?

☐ Vanilla   ☒ Mint

☐ Strawberry   ☐ Toffee

☒ Chocolate

What do you like best?

☒ Cone   ☐ Tub

[   ] Other

2) Look at the tally chart. It shows a tally of favourite sports.

Sport	Tally
⚽	ЖЖ ЖЖ
🎾	|||
🏀	ЖЖ
🏊	||||

   a) Which sport is the *most* popular?
   b) How many people like basketball?
   c) How many people like swimming?
   d) Which sport is the *least* favourite?

3) Ask 10 people what is their favourite sport and record your answers in a tally chart.

Look at the chart that Isla created to show the eye colours in her class.

Hazel	Blue	Green	Brown
✓	✓	✓	✓
✓	✓	✓	✓
✓	✓	✓	✓
✓	✓		✓
			✓
			✓
			✓

1) Put this data into a table with tally marks and show totals.

2) With a partner collect your own data and display it in the same way.

14.2 Bar graphs

We are learning to create bar graphs.

Before we start

Find out what pets the people in your class have. Record the data you collect using tally marks.

Data can be displayed in graphs and charts.

Let's learn

A **bar chart** is a type of graph where the data is represented in bars.

This **bar chart** shows the eye colours in Primary 3.

It shows four children have hazel eyes, six have blue eyes, two have green eyes and eight have brown eyes.

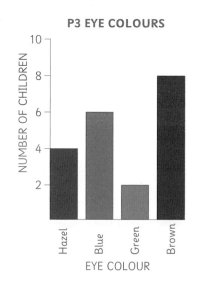

P3 EYE COLOURS

NUMBER OF CHILDREN

EYE COLOUR

Hazel · Blue · Green · Brown

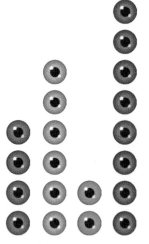

This is a **pictograph** displaying the information.

Bar graphs and pictographs can display the information **vertically** or **horizontally**.

1) Read the information in the box. Complete the vertical picture graph.

Use this symbol: ☺

Four students like the movie *Frozen*.

Six students like the movie *Cars*.

Eight students like the movie *Brave*.

Seven students like the movie *Toy Story*.

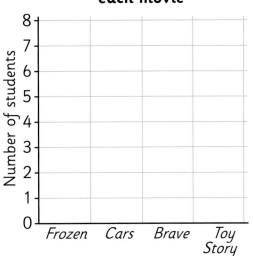

Number of students who liked each movie

Resource 1B_14.2_Let's Practise_Q1

CHALLENGE!

Complete the totals on the data collection sheet. Use this information to draw a bar graph in your jotter. Remember to include a title and labels on your bar graph.

Survey question: What is your favourite land vehicle?

Data collection sheet		
Land vehicle	**Students who liked this vehicle**	**Total**
Motorbike	////	
Tractor	///	
Truck	///////	
Racing car	//////////	

14.3 Investigate and interpret information

We are learning to investigate and interpret information.

Before we start

Number of students who like each shape

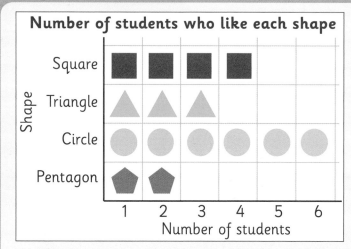

1) Which shape is the most popular?

2) Which shape is the least popular?

3) How many students were asked altogether?

Graphs and charts tell us information.

Let's learn

We see graphs and charts everywhere.

They are on the internet, television, and in books and newspapers.

We can say what the graph is telling us if we have the skills to read and interpret the data.

1) Graph A shows the results of a survey, 'What is your favourite wild animal?' Look at the graph and answer the questions.

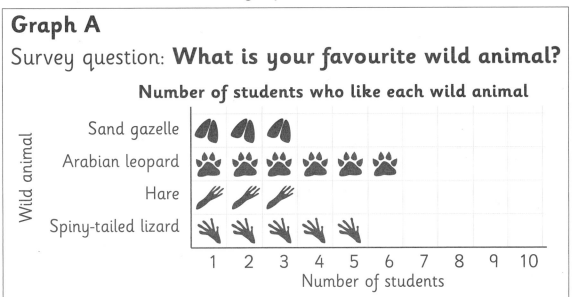

Graph A

Survey question: **What is your favourite wild animal?**

Number of students who like each wild animal

a) How many students chose the sand gazelle?

b) How many students chose the Arabian leopard?

c) Which wild animal received five votes?

d) Which two wild animals received the same number of votes?

2) Graph B shows the results of a survey, 'What is your favourite woodwind instrument?' Look at the graph and answer the following questions:

a) Which instrument received the most votes?

b) Which instrument received the fewest votes?

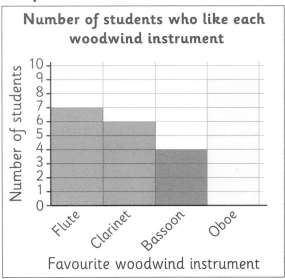

Graph B

Number of students who like each woodwind instrument

c) Which instrument was more popular than the flute?

d) The clarinet was more popular than the ___**?**___

but not as popular as the ___**?**___ and

the ___**?**___ .

3) Which graph does each statement refer to: Graph A or Graph B? Write the letter of the correct graph that relates to each statement in your jotter.

a) The bassoon received the fewest votes.

b) The highest scoring category received nine votes.

c) The lowest scoring category received three votes.

CHALLENGE!

Look at this graph. It shows the results of a survey about how students get to school.

Write four statements about this graph in your jotter.
For example:

'I notice that three more students walk to school than come to school by scooter.'

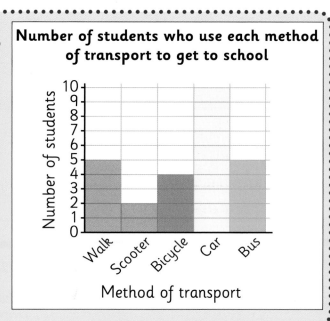

Number of students who use each method of transport to get to school

14 Data handling and analysis

14.4 Drawing conclusions from data and graphs

We are learning to draw conclusions from data and graphs.

Before we start

Look at the graph and discuss with a partner what information you can see.

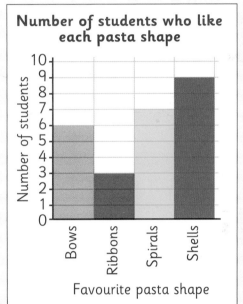

Number of students who like each pasta shape

Number of students

Bows | Ribbons | Spirals | Shells

Favourite pasta shape

Data and graphs can tell us lots of different things.

Let's learn

Data and graphs help us to draw conclusions and understand information we collect in a survey.

Always look carefully at the data and graphs to make sure you understand them before you answer any questions.

Let's practise

1) Use the data collection sheet to collect data from students for a survey.

Survey question: **What is your favourite percussion instrument?**

Resource 1B_14.4_Let's Practise_Q1

Percussion instrument	Number of students
Triangle	
Tambourine	
Drum	
Cymbals	

2) Construct a bar graph to show the results of your survey.

Resource 1B_14.4_Let's_ Practise_Q2

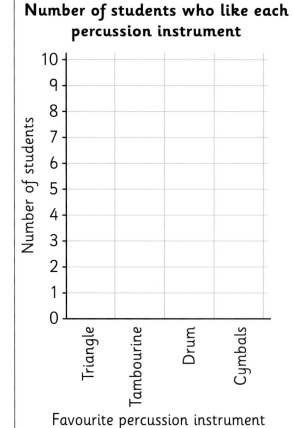

Number of students who like each percussion instrument

Number of students: 0 1 2 3 4 5 6 7 8 9 10

Triangle, Tambourine, Drum, Cymbals

Favourite percussion instrument

3) Answer these questions about your bar graph:
   a) How many students chose the cymbals?
   b) How many students chose the triangle?
   c) What was the most popular percussion instrument?
   d) Which instrument received the fewest votes?
   e) Pick two of the percussion instruments. How many more votes did one instrument receive?
   f) Did any instruments receive the same number of votes?

CHALLENGE! ....

Resource 1B_14.4_ Challenge

1) Look at the data Isla has collected.
   a) What conclusions can you draw about students reading in her class?
   b) Use the data to create a bar chart.

Five students read one book.

Eight students read two books.

Three students read three books.

Four students did not read a book.

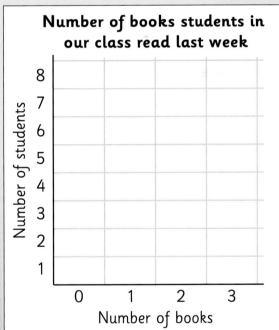

Number of books students in our class read last week

Number of students (y-axis: 1 to 8)

Number of books (x-axis: 0, 1, 2, 3)

15 Ideas of chance and uncertainty

15.1 Understanding chance and uncertainty

> We are learning to understand chance and uncertainty.

Before we start

With a partner, find an example of something that is:
- Certain
- Possible
- Impossible

> Things can be **certain**, **possible**, **unlikely** or **impossible**.

Let's learn

We use the following words to describe the chance of something happening:

- **Impossible** — it will never happen

- **Unlikely** — it may happen but probably will not

- **Likely** — it may happen and probably will

- **Certain** — it will definitely happen

Events can be placed on a **probability scale** to show the chances of them happening:

Impossible　　　**Unlikely**　　　**Possible**　　　**Certain**

1) Where on the **probability scale** would the following events be placed?

a) An elephant will drive the school bus.
b) I will play outside today.
c) The Queen will visit the school.
d) I will be older next week.
e) It will snow on Christmas Day.
f) I will eat something today.
g) It will get dark tonight
h) Humans will travel to Mars on a rocket.

2) With a partner, give an example of something that is: *impossible*, *unlikely*, *possible* and *certain*.

3) Answer impossible, unlikely, possible or certain:

a) I will roll a six.
b) I will meet an alien.
c) I will toss a coin and it will land on heads.
d) Frogs will fall from the sky.

CHALLENGE! ....

Impossible   **Unlikely**   **Possible**   **Certain**

Copy the probability scale and draw two events on each section.

Share your drawings with a partner.

Does your partner agree with where you have put the events, or do you need to move them?